PREHISTORIC LIFE

Text by Beverly Halstead
Illustrations by Jenny Halstead

HarperCollins*Publishers*

HarperCollins Publishers
PO Box, Glasgow, G4 0NB

First published 1989

Reprint 10 9 8 7 6 5 4 3

ISBN 0 00 458819 3

For Tom

Printed in Great Britain by
HarperCollins Manufacturing, Glasgow

Contents

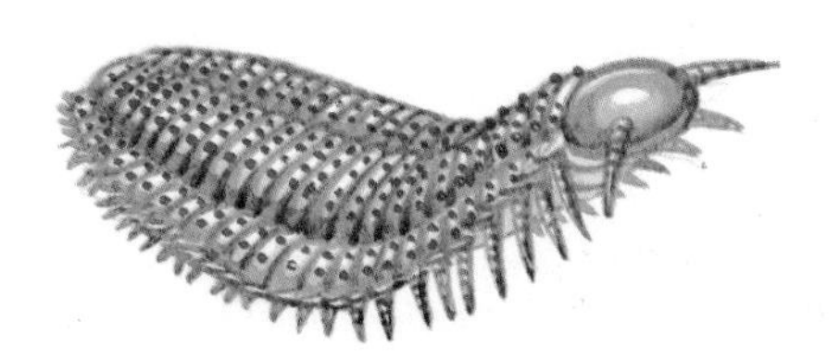

Introduction

Dinosaurs ruled the Earth for 140 million years. That is nearly thirty times as long as the entire history of humankind from the time when human beings first separated from the apes about four to five million years ago. Modern humans have existed for only 200,000 years; the dinosaurs lasted 700 times longer, or to put it another way, 14,000 times the time since man's first civilisations.

Knowledge of these monstrous denizens of the past comes from fossilized bones and teeth. Early naturalists were puzzled by these curious stony shapes often embedded in hard rocks. Some thought them meteorites or gemstones, although in 1666 a Danish scientist called Niels Stensen noticed the similarity of some with shark's teeth and concluded that they were petrified teeth. His English contemporary, Robert Hooke also considered them organic remains. For a time they were believed to belong to creatures which died in the biblical Flood but gradually scientists came to understand that fossils were preserved in many different strata of rock, laid down, one upon another, over very long periods of time.

About a century and a half ago scientists began to realise that the Earth had once been inhabited by giant reptiles on the land, flying reptiles in the skies and swimming reptiles in the seas. The British anatomist Sir Richard Owen studied fossils and

Mamenchisaurus
(*see page 101*)

attempted reconstructions of animals from fossilized remains. In 1841 he invented the name Dinosauria for the huge plant-eating and flesh-eating reptiles which captured the imagination of the Victorian public as they do our own.

During the past 200 years millions of fossils have been collected. From their study the overall pattern of the evolution of life on Earth has been established. We now known that the plates of the Earth's crust continually move. Their collisiön pushes up mountains. By 250 million years ago the continents had come together to form a single super-continent, Pangea, then broke away to form the modern continents. Successive Ice Ages have also changed climates and sea levels. A fossil found on a northern hillside may once have been in a tropical sea.

The record of life on Earth goes back nearly 4,000 million years, for more than half of which the planet was inhabited only by bacteria. Then, about 1,500 million years ago more complex organisms began to develop until at last there came the dinosaurs and, eventually, man. In this book you can find the whole story from the beginning of life to the birth of civilisation, a mere 10,000 years ago.

The Fossil Record

The word fossil originally meant anything that was dug up (Latin *fossere* = to dig) but has now come to mean the preserved remains of past life. To be dug up a fossil must first have been buried, not in a hole but by being covered. This is how it happens.

The formation of sedimentary rocks
The rocks that make up land are worn by wind, rain and the action of plant roots and gradually disintegrate, rain actually dissolving certain minerals. This is the first stage of erosion (Latin *rodere* = to gnaw), or eating away, of the surface of the land. Streams and rivers carry away the rock debris, transporting it towards the sea. Larger pieces of rock are rolled and

dragged along the river bed, sand grains are bounced along, flakes of clay are carried in a muddy suspension and some minerals are dissolved and carried in solution. An animal that falls into a river will float downstream.

When the river reaches the sea it can no longer carry its load of minerals and they are dropped as sediments (Latin *sedere* = to sit or settle). Gravels and sands are the result of the flow of current stopping. Clay particles settle after contact with salt in the water. Dissolved minerals are extracted by animals and plants to make skeletons which later accumulate on the sea floor. The deposition of sediments covers the remains of animals and thus can preserve them as future fossils. The process of erosion, transport and deposition may take months or years and the accumulating sediments can lie undisturbed for millions of years.

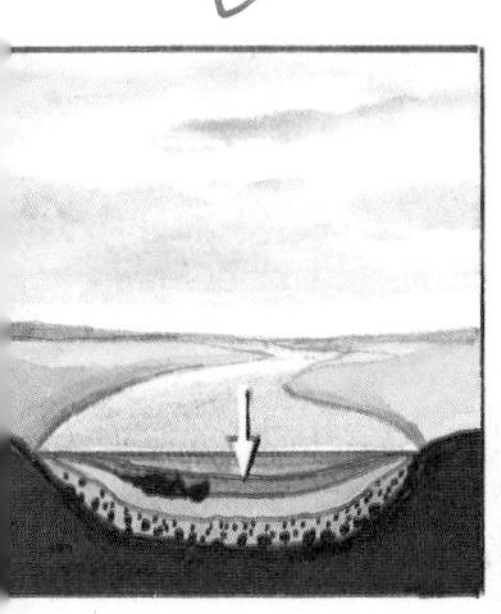

The formation of fossils

When an animal or plant dies it is usually eaten by insects and worms, finally fungi and bacteria break it down completely and its constituent chemicals are recycled. The chances of a land-living animal becoming preserved in sediments are remote. The land is the source of most sediments not the place where deposition takes place. There are exceptions, animals that fall into rivers may be preserved in river sediments, animals may be engulfed in sand storms, or be trapped in caves. Occasionally special conditions favour the formation of fossils: where a flying bat falls into a lake, and drowns, it will become water-logged and sink to the bottom. If muds and sands are being carried into the lake, they will cover the body and the skeleton will become preserved.

Sometimes bones, or even the soft parts of animals and plants are dissolved away and replaced by other minerals, leaving a cast of their shape. If skilled palaeontologists (experts in the study of such a fossil hole unfilled they may be ab with rubber or plastic to produce a replica.

The shells of shellfish that inhabit coastal sa muds are calcium carbonate, the compon limestone. They are already buried in sedim cannot avoid becoming fossils and make up majority of common ones. The richest deposits where storms have swept shells into concentrat banks. Away from coasts the shells of microscop organisms rain down continually onto the ocean floor, only there is a complete record preserved.

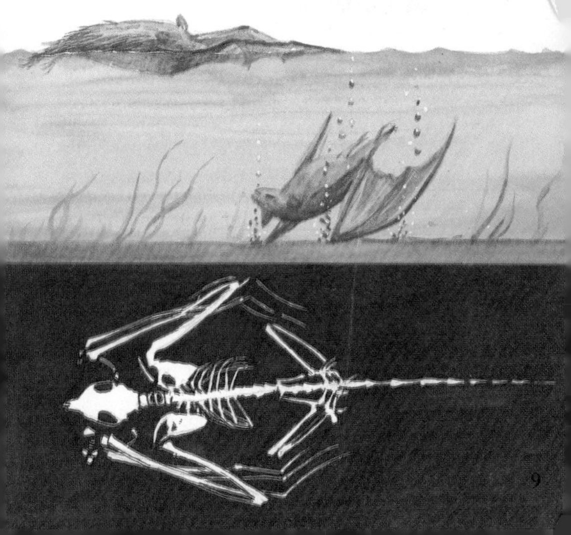

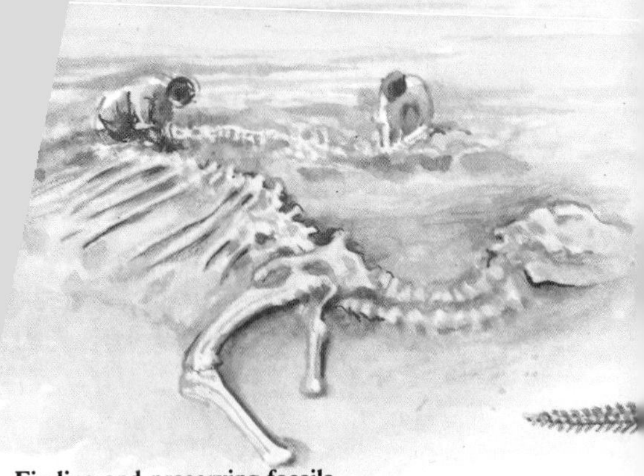

Finding and preserving fossils

The record of the rocks cannot be read until earth movements have raised the sediments to form dry land and they have been eroded to expose the different layers.

The next stage depends on luck. Sometimes fossils are found by miners cutting or splitting coal or rock; sometimes they are exposed by weathering, but then they must be found before the wind and rain destroy them forever. Certain layers become famous for their fossils and are carefully searched. Some rocks are made up entirely of fossils, every scrape of Chalk will fill one's finger nails with thousands of microscopic fossils.

When dinosaurs, or other fossils, are discovered, the bones are carefully exposed and coated with protective layers. Then, together with the surrounding sediments they are transported to a laboratory for the fossils to be extracted. Air abrasives, dental and percussion drills and weak acids are all used to reveal their details. After careful study groups of bones which make up an animal may be mounted in their life positions and put on display in a museum.

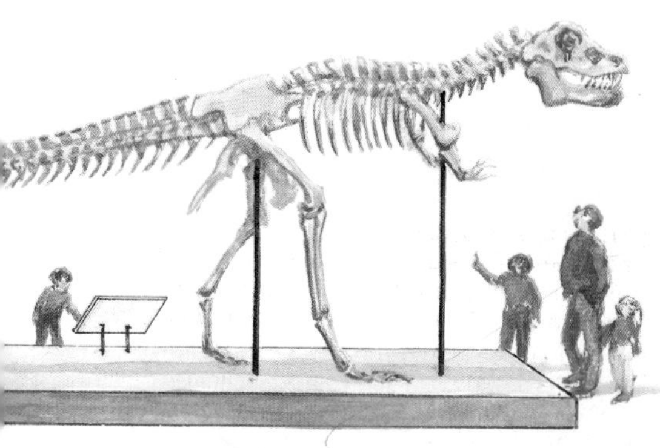

Geological time

The fossil record is divided into geological periods, often named after the places where they were first described: the Devonian from Devon, Permian from Perm in Russia and Jurassic from the Jura mountains of France. The later sediments are always laid down on top of the older and rocks can be dated by

TIME CHART

			beginning date in million years ago
Caenozoic	Quaternary	Holocene	.01
		Pleistocene	1.6
	Tertiary	Pliocene	5.3
		Miocene	23
		Oligocene	34
		Eocene	53
		Palaeocene	65
Mesozoic	Cretaceous		135
	Jurassic		205
	Triassic		250
Palaeozoic	Permian		300
	Carboniferous		355
	Devonian		410
	Silurian		438
	Ordovician		510
	Cambrian		570
Precambrian			4500

measuring the decay of radioactive substances, such as uranium, which over millions of years changes into a type of lead. From the relative amounts of these elements remaining, radiometric dates in years can be given for all the geological periods.

These are the dates (in millions of years ago) when various lifeforms first appeared, with the comparative timescale clearly shown in the diagram *right*:

1 Bacteria -3,800 million years
2 Blue-green algae or cynobacteria -2,900, first organisms to use the energy of sunlight
3 Eukaryotes -1,450, first true animal and plant cells with a nucleus

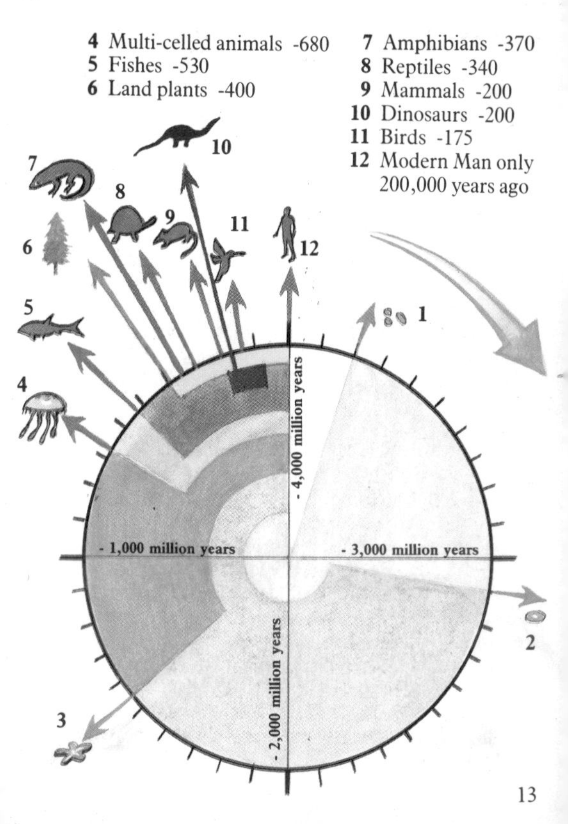
4 Multi-celled animals -680
5 Fishes -530
6 Land plants -400
7 Amphibians -370
8 Reptiles -340
9 Mammals -200
10 Dinosaurs -200
11 Birds -175
12 Modern Man only 200,000 years ago
10
7
8
9
11
6
12
5
1
4
- 4,000 million years
- 1,000 million years
- 3,000 million years
- 2,000 million years
2
3

Classification

The study of fossils has given a great deal of information about the evolution of forms and helped to explain the relationships between living things. This makes it possible to classify them in a system of successive groups within groups (*diagram right*). The method that is used for all living things was devised by Carl Linnaeus (1707-1778) in his *Systema Natura*. Discoveries are still being made so that at the most detailed level a new relationship may be found, or a plant or animal thought to be a new find turns out to be identical with one that had already been named. This is why scientific names are still occasionally changed.

The first major divisions are into kingdoms such as the animal and plant kingdoms.

Within a kingdom are separate branches, the phyla, such as the segmented worms, the echinoderms (starfish and their relatives), arthropods (jointed limbed animals such as crabs), molluscs (snails and their relatives) and the chordates (sea squirts, man and other backboned animals).

Phyla can be divided into classes, for example fish, amphibians, reptiles, birds and mammals. Within the mammals there are about twenty orders, such as carnivores, even-toed ungulates and the Primates, to which the lemur, ape and human families belong.

Families include genera, which are always written in italics, such as *Homo* (man). Within a genus, there may be several species or only one as for example with modern man *Homo sapiens*.

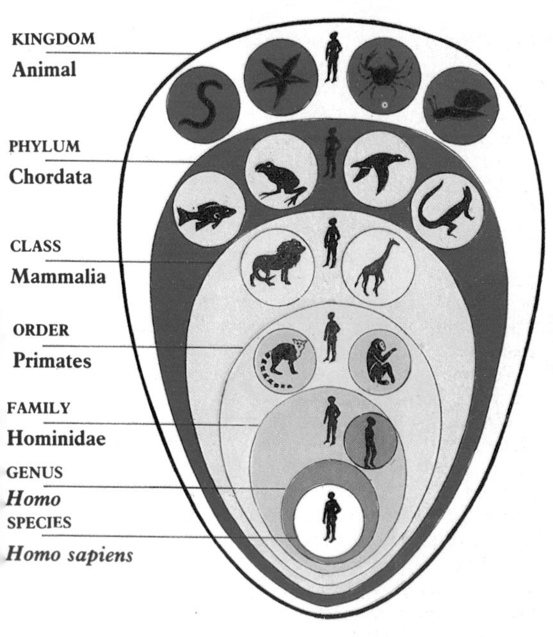

The way in which all living things can be classified in a pattern of groups within groups is one of the evidences of evolution, for the arrangement only makes sense if it is the result of the descent of different kinds from common ancestors.

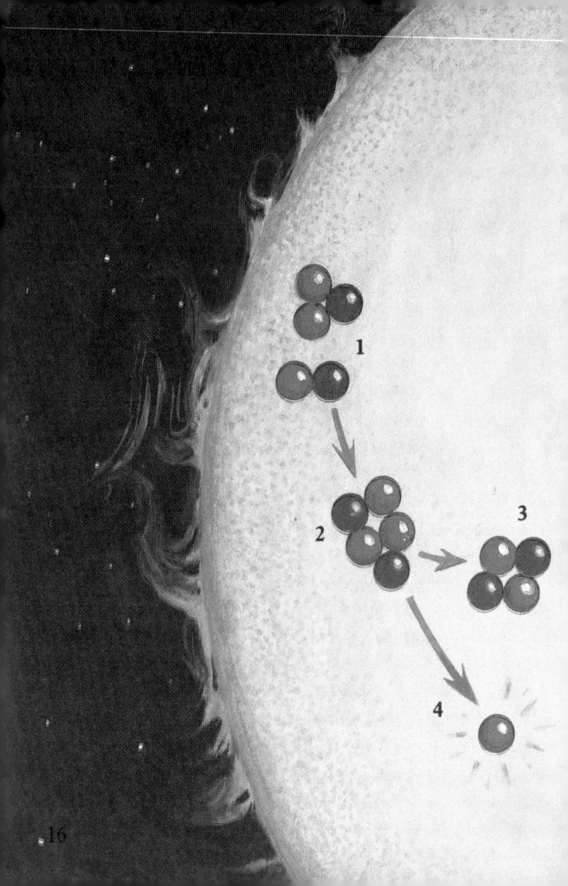
1
2
3
4

The Origin of Life

The universe is made up almost entirely of the light gas hydrogen in empty space. When a cloud of hydrogen (**1**) is sufficiently large, the force of gravity – the attraction that all the particles exert upon each other – becomes so great that the molecules are fused together (**2**) to create the heavier element helium (**3**). Helium is lighter than the amount of hydrogen that made it. The difference has been converted into energy (**4**). Every second, some 600,000 tonnes of the Sun's hydrogen is fused into helium. The energy, released into space, floods the Earth.

Deep within the furnace of the sun enormous pressures produce further fusion which creates heavier and heavier elements, such as oxygen, iron, nitrogen, sulphur, carbon and silicon, up to the heaviest, uranium.

The planets which revolve around the Sun are composed of oxygen, iron, nickel, carbon and silicon as well as hydrogen. These elements are combined into a number of simple compounds, such as water, methane, silica and carbon dioxide. They are the debris from the inner furnace of a now extinct star - a long lost companion of the Sun.

The destruction of one Sun provided the habitat for life; a still living Sun provides the energy to maintain life - but not all planets can support life. Around every star there is a zone, known as the

Saturn Jupiter

ecosphere, where life might be possible. Its temperature is such that water can exist in its liquid phase – a necessity for all processes of life are performed in aqueous conditions. There must also be a solid body large enough to have sufficient force of gravity to hold on to water in its liquid state, and to hold gases such as carbon dioxide.

In our own star system, Mercury is too small and too close to the Sun. Venus, Earth and Mars all come within the ecosphere. Beyond them a belt of asteroids contains rocky lumps, some of nearly moon size but too small to hold an atmosphere. The outer planets such as the giants Jupiter and Saturn are composed of gases and cannot support life as we know it.

Within the ecosphere the surface of Venus has a temperature of about 750°C (1382°F), with clouds about 475°C (887°F). Its atmosphere is made up of water vapour, carbon dioxide and sulphur dioxide. In part of the clouds, reactions seem to take place in the liquid phase. It is possible that certain sulphur bacteria, such as those that live on Earth in hot

Asteroids Ecosphere Mercury Sun

springs (*see page 22*) could live in the clouds of Venus. Seeding of the clouds of Venus could change the composition of its atmosphere and make Venus habitable for more advanced living things.

Mars has a thin atmosphere of carbon dioxide but water is frozen; yet there is evidence that in the past there was running water. This means that life could have existed on Mars during such times and, with appropriate technology, the planet could support life in the future.

Only Earth has the right conditions to keep water in its liquid state and to hold on to the gas carbon

Mars **Earth** **Venus**

dioxide, although the composition of the atmosphere has now been changed to one of oxygen and nitrogen with only a small amount of carbon dioxide.

The building blocks of life

The original atmosphere of the planet Earth was like the gases that come out of volcanoes: mainly water (**1**) and carbon, carbon dioxide (**2**) with other simple compounds such as methane (**3**) and ammonia (**4**). Ultraviolet radiation from the Sun as well as electrical storms in the clouds of gas resulted in the formation of simple aminoacids, such as glycine (**5**). Aminoacids can be produced in a laboratory by passing an electrical current through a mixture of gases. They are the building blocks of proteins, which form the fundamental structure of all living things.

In dry heat, like that found on the cooling surfaces of molten lava, aminoacids can be fused together to form simple proteins. When water vapour in the clouds of volcanic gases condenses, rain will fall and douse the hot rocks. The primitive proteins form tiny globules which have a simple membrane. Within these globules simple chemical processes can take place. This seems to be the very beginning of life. The essence of life, however, is that such structures must be capable of processing the outside environment without substantially being altered themselves, rather like the catalysts of chemical reactions. Exactly how this began is still not understood but the chemical activities in these globules suggest how life may have started.

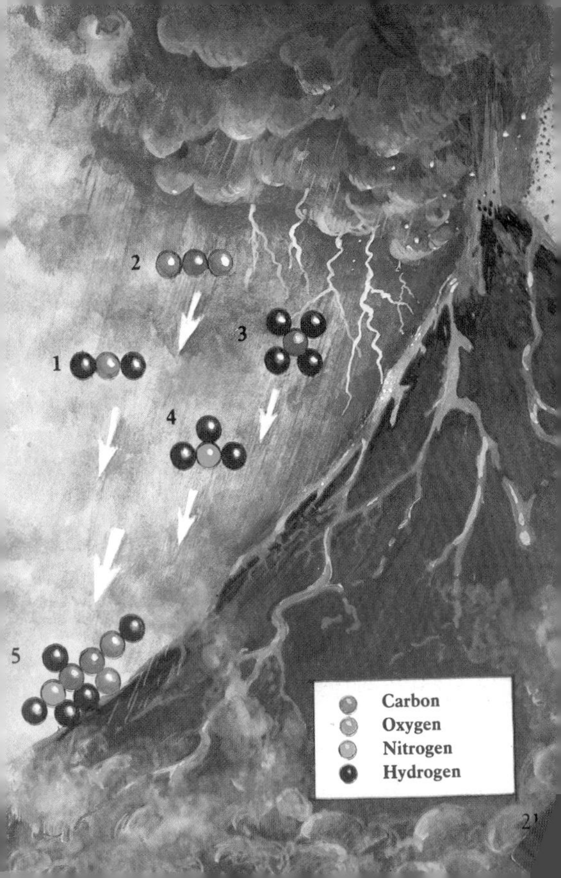
2
3
1
4
5
Carbon
Oxygen
Nitrogen
Hydrogen

The Precambrian Era
(4500-570 million years ago)

The first life

The first signs of life on Earth are tiny globules from rocks 3800 million years old, the most primitive of all bacteria (**1**). The first definite living things, they obtained their energy for life by breaking down molecules, such as sulphur compounds, associated with volcanic activity on the sea floor (**2**). Then, some 3000 million years ago, bacteria became able to obtain their energy directly from sunlight (**3**). They could also store nitrogen, exactly as soil bacteria do today.

A major stage in the history of life was the appearance of cyanobacteria (cyan = blue) or

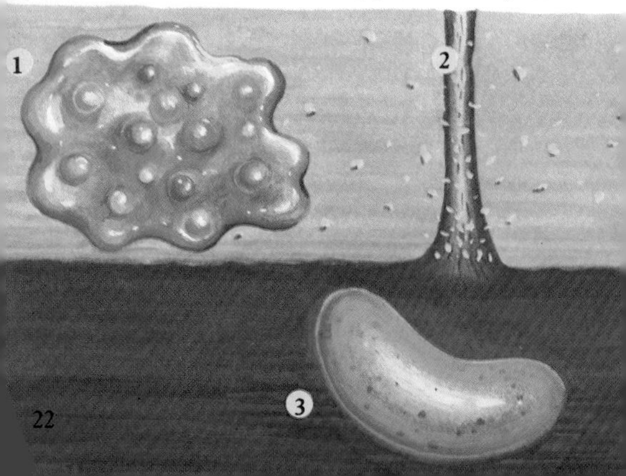

blue-green algae (**4**), which formed mats and sometimes cabbage-like mounds on the sea floor, as seen today off the coast of Australia (**5**). These algae used the energy of sunlight to combine water and carbon dioxide to make carbohydrates. This process, called photosynthesis (= light-combining), gives off oxygen as a waste product. Some of the oxygen combined with iron and was deposited on the sea bed as banded ironstone formations, a 'rusting of the oceans' which happened 2500-1750 million years ago.

The algal mats also incorporated layers of fine muds and calcium carbonate so that layers of limestone were formed, known as stromatolites or stone blankets. Once the iron had been deposited, free oxygen began to accumulate in the waters of the oceans and also to be released into the atmosphere where it began to form an ozone layer which provided a shield against the dangerously strong ultraviolet radiation from the sun.

Bacteria of all kinds

For 2000 million years bacteria were the only living things on the Earth. The earliest were the heat resistant bacteria (**1**), now found in sulphur springs. Next came nitrogen-fixing bacteria (**2**), now found in soils. There were also spirochaetes (**3**), capable of active movement, with an internal structure consisting of a ring of nine pairs of tubules with an extra pair in the centre. The most advanced were the cyanobacteria (**4**) which produced free oxygen.

Most primitive bacteria can only flourish in environments depleted in oxygen – to them it is a dangerous poison. The great advance in the evolution of these micro-organisms came when they were able

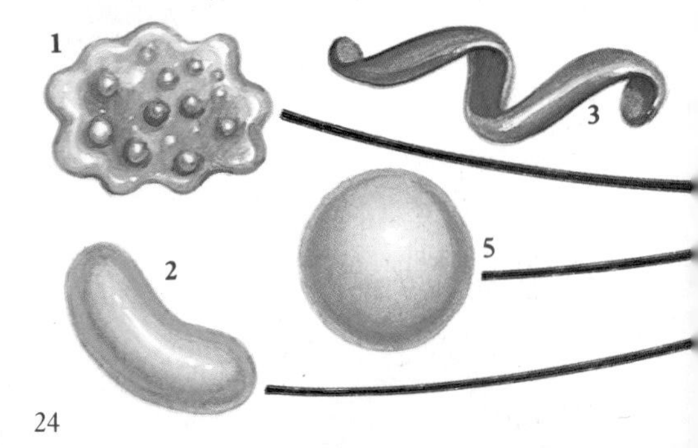

to use oxygen as a fuel to burn complex compounds as food for the release of energy.

Single-celled plants and animals

About 1500 million years ago there was a dramatic increase in the size of micro-organisms. This marked another crucial stage in the evolution of life. Different types of bacteria combined with a nucleus bacteria (**5**) to make up the first single-celled animals (**6**), which were able to use oxygen. When these animals enclosed photosynthesising cyanobacteria, the first single-celled plants (**7**) were formed, which could use sunlight to synthesise food and oxygen to break it down.

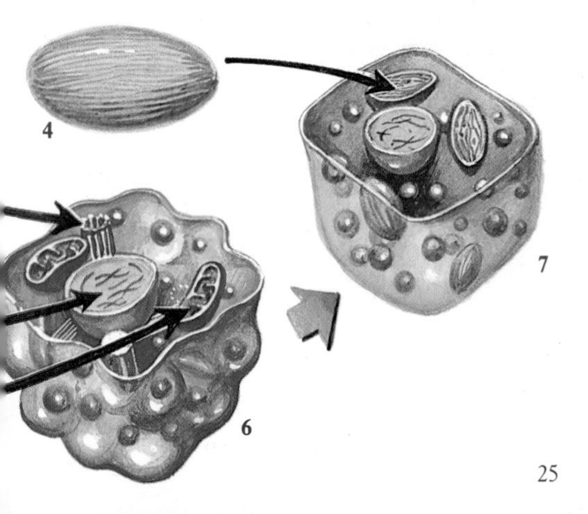

The simple animal and plant cell is the result of several different types of living thing combining so that each type turned into a miniature organ (or organelle) within the cell.

The living cell

All animal and plant cells have the same basic internal structures. The genetic material deoxyribonucleic acid (DNA), through which the characteristics of life forms are passed on, is contained in the nucleus, or kernel (**1**). The cell also contains a number of miniature organs or organelles.

One of the key organelles is the mitochondrion (**2**) which regulates the energy of the cell. It has its own DNA and seems to be derived from a soil-type bacterium. In plant cells the organelles concerned with photosynthesis, the chloroplasts, also have their own DNA. These are cyanobacteria that have been incorporated into a basic animal cell to create the first plants. Other parts of the cell, all derived from spirochaetes, include the centrioles (**3**), which organise the division of the cell, and the flagellae (**4**) and cilia (**5**), which set up currents and move cells.

The internal structure of the organelles is identical to different types of bacteria, which means that the basic animal and plant cell is the result of these different organisms combining to form a cooperative whole. The fossil record indicates that this took place 1500 million years ago. This major leap forward in evolution was accomplished not by competition but by cooperation.

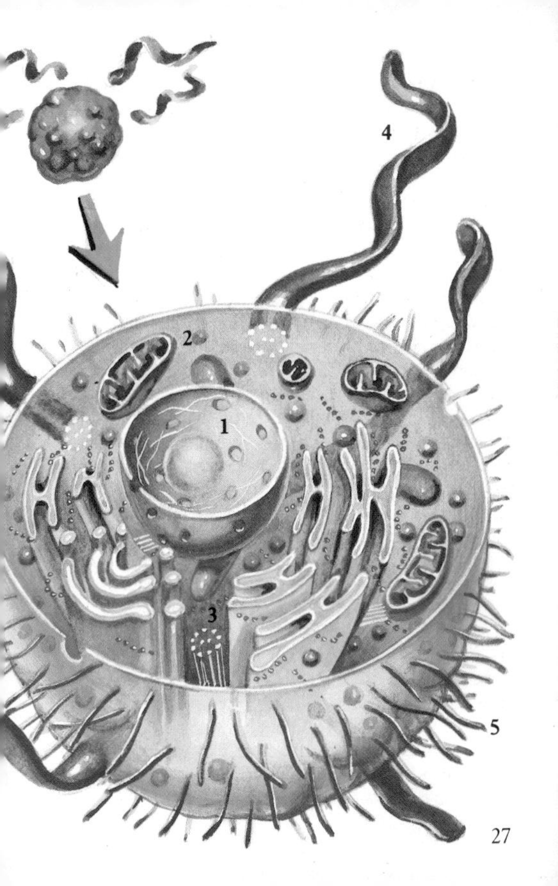
4
2
1
3
5

The origin of sex

The speeding up of the evolution of life on Earth was due entirely to the invention of sex. A means of mixing up the hereditary material of two individuals, so that the offspring necessarily differed from their parents. In every animal and plant cell, the DNA is arranged in pairs of chromosomes (**1**) in the nucleus. The chromosomes carry the genes which determine individual characteristics.

The chromosomes are released from the nucleus and arranged on the spindles formed from the centrioles (**2**). Then the cells divide (**3**) to produce the sex cells (**4**), with only one set of chromosomes. Those that have a large food supply, or yolk, are known as eggs; those that are tiny but very mobile are sperm. When the sex cells combine (**5**) the paired condition is restored (**6**). This process increases the variation among individuals. Those that were better adapted to the environment stand a greater chance of survival. Natural selection of such individuals is the main process of evolution.

The origin of sexual reproduction seems to be connected with primitive organisms ingesting one another. The genetic material of two individuals became combined so that the fused organism had two sets. A large organism was now produced which then split in two so that each new organism had a single set of genetic material. There began an alternation between single and paired sets, with eventually the paired condition becoming the normal situation for both animals and plants.

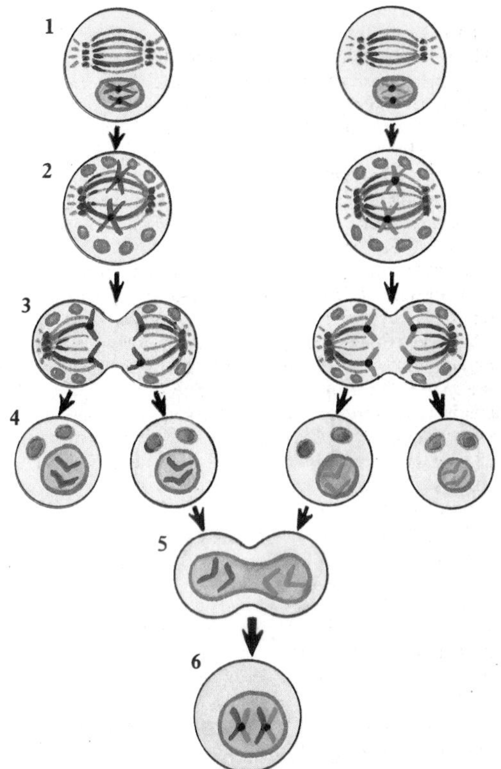
1
2
3
4
5
6

The first multi-celled animals

The next great advance in the history of life, after the development of sexual reproduction, was the evolution of organisms made up of many cells, with different types of cells specialised for specific tasks.

By about 750 million years ago there were large soft bodied animals whose fossils are found preserved as impressions in tidal sands, which can be identified by a symmetrically rippled surface. These fossils are found in Precambrian rocks in England, Wales, Namibia, Siberia and Australia, and are known as Ediacara animals from Ediacara in Australia. Among them are many kinds of circular fossil, which seem to have been forms of jellyfish (**1**) which floated in the seas capturing minute organisms with their tentacles. Fixed to the sea bed there were frond-like sea-pens ***Charnia*** (**2**) and *Xenusion* which were related to

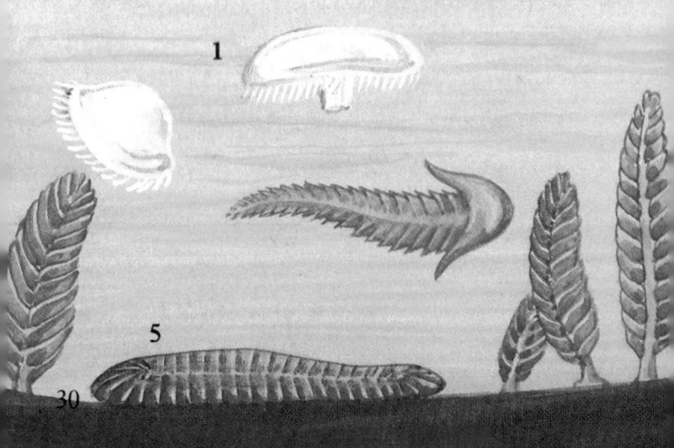

corals, and fed by capturing small organisms in their tentacles. Burrowing in the sands were worms (**4**) which must have filtered minute food particles from the water.

There were flat animals that looked segmented and may have been a type of worm (**5**) which crawled over the bottom feeding on mats of algae. One form ***Spriggina*** (**3**) was flat and segmented but had a distinct head; this was possibly a swimming animal related to the arthropods (the spider and lobster family) or jointed-legged animals such as crabs and insects.

Among all these animals there were no active carnivores, all fed on microscopic organisms. Many of the Ediacaran animals cannot be directly linked to known living animals, but they seem to represent an important stage in the history of life.

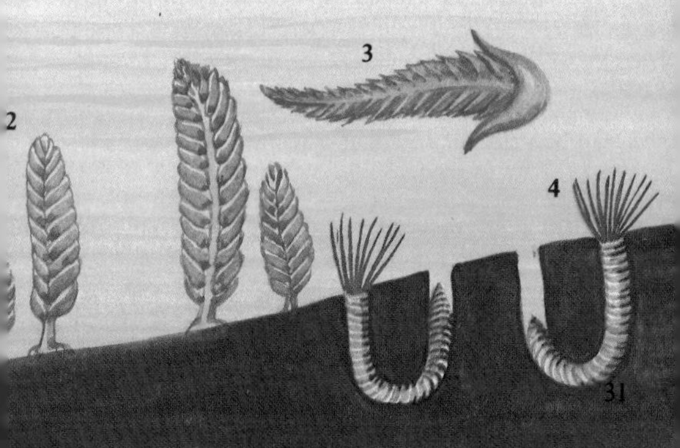

Hydrothermal vents

Most life on Earth is dependent on the Sun for its energy, but life can develop wherever energy exists. Volcanic heat and gases come from within the Earth and are a source of energy absolutely independent of the Sun.

Along the floors of all the oceans there are cracks in the crust, known as hydrothermal vents, through which hot gases pour into the cold water of the abyssal depths. These gases emerge at 300°C (572°F) and the sudden cooling as they hit water at 4°C (39.2°F) results in sulphur compounds and metallic ores forming tall chimneys several metres high on the sea floor. Around them is a narrow zone of a few metres where bacteria flourish using sulphur compounds as their source of energy.

The bacteria form the base of a complex food chain confined to the region of hydrothermal vents. Today these pockets of life contain groves of giant tube worms (**1**), clams filtering bacteria (**2**), grazing worms (**3**), small crabs (**4**) and curious squids that feed on the crabs and the other molluscs. All the types of animals associated with the vents are new to science, but they can be classified with living groups. The significance of the hydrothermal vent animals is that they represent an ecosystem on Earth that is not dependent on anything that goes on in the rest of the planet. It is theoretically possible, therefore, for life to develop wherever the fundamental requirements are present: an energy source and the physical parameters allowing water to exist as a liquid.

1
2
3
4

The Palaeozoic Era
(570 - 250 million years ago)

Fossils suggest that at the beginning of the Cambrian period, 570 million years ago, all kinds of animal life seemed suddenly to appear. What actually happened is that many groups of organism began to produce skeletons of calcium salts, and these were preserved in the rocks in great numbers, thus giving the misleading impression of a dramatic explosion of life on Earth.

Siberian buttons

The first signs of skeletons are microscopic button-like shapes discovered in Siberia in rocks of both Precambrian and Cambrian age. They are now known from Turkey, Spain, Greenland, Spitsbergen, Kirgiz and Estonia. These buttons are made of calcium phosphate, the mineral of bones and teeth. They seem to have been embedded in the skin with their outer surface ornamented with minute protruding tubercles.

The circular buttons with tiny tubercles are named ***Lenargyrion knappologicum*** (**1**, **2**) which means the little coin from the Lena river of Siberia, knappology is the study of buttons. The forms with large tubercles are ***Hadimopanella*** (**3**, **5**) and the oval buttons ***Kaimenella*** (**4**). It is just possible that these are the the remains of the first vertebrates but nobody really knows to what sort of animal they belonged.

1
2
3
4
5

The origin of skeletons

Certain substances such as salt and calcium (**1**) soak into the cells and tissues of organisms living in the sea, which must get rid of them. The organelle concerned with energy exchange, the mitochondrion (**2**), soaks up calcium and phosphate and pumps it out as calcium phosphate. This mineral can be deposited within cells (**3**) or completely removed from them and built into an internal skeleton, as in single-celled animals (**4**) and vertebrates, or an external one as in arthropods.

The laying down of a skeleton is usually by seeding mineral crystals on to the fibrous protein, collagen, which is the main structural protein of the animal kingdom, and makes up about a third of all the protein of vertebrates. Collagen can only be formed in the presence of free oxygen, and the appearance of collagen and skeletons seems to have taken place

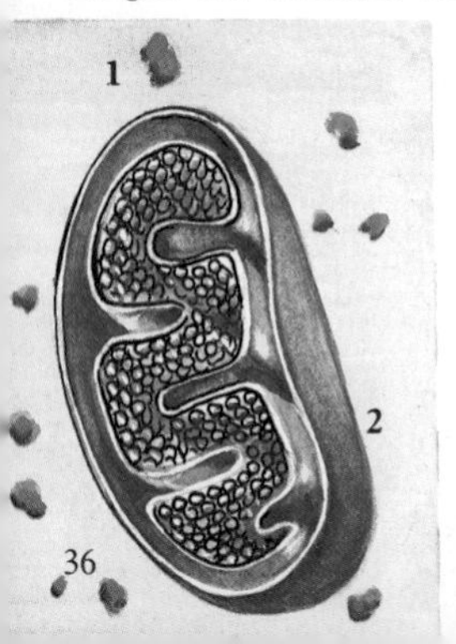

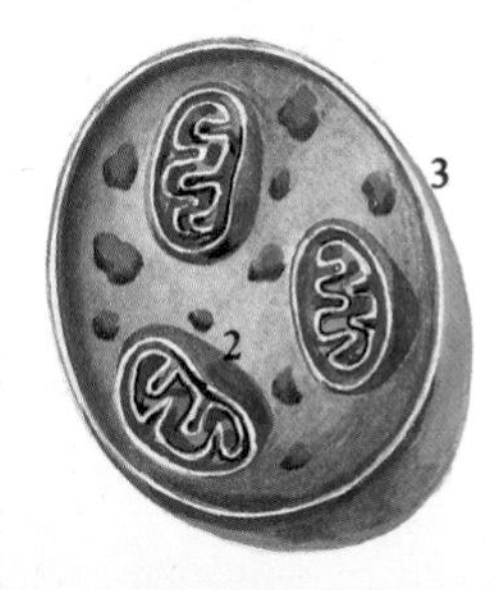

when there was 1% of free oxygen in the atmosphere. The combination of the ability of the mitochondria to concentrate calcium, the formation of collagen and the excess of calcium in sea water resulted in the creation of skeletons.

The Burgess Shale

The fossil record is dominated by animals and plants with mineralized hard parts, but there are very rare situations where all the soft-bodied animals are also preserved. The most famous example is in British Columbia, Canada, the Burgess Shale of Cambrian age (570-510 million years ago). This rock was formed from very fine muds, which accumulated on the ledges of a huge underwater limestone cliff. Periodically a slurry of mud would slip off a ledge and fall to the bottom of the cliff, carrying with it all the animals. As the muds accumulated the animals would be

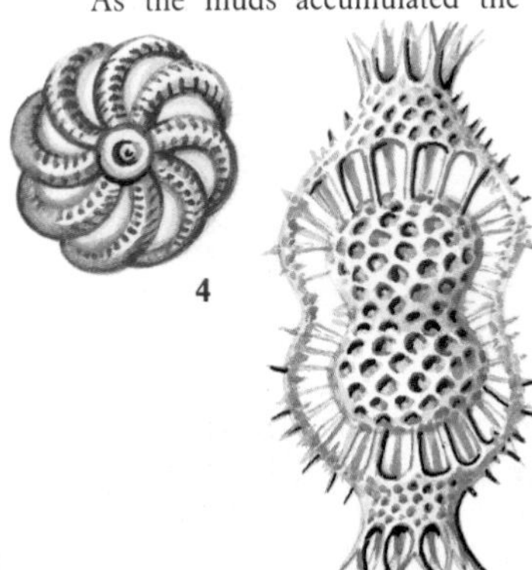

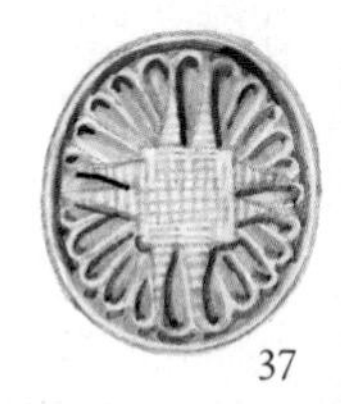

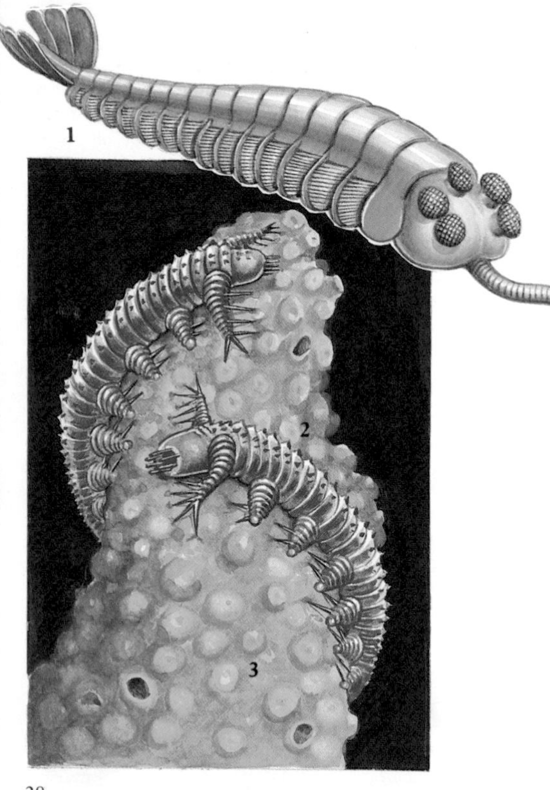
1
2
3

compressed by the weight above, but because the water was so deep there were no other animals to disturb or destroy the remains. Many kinds of worms burrowed in the mud of the ledges, other animals crawled over their surface and swam in the water above. The most dominant group were the arthropods, which made up 43% of the known species, including crustaceans such as fairy shrimps.

The thousands of fossils in the Burgess Shale are preserved in fine detail – even delicate filaments of gills survive intact, as a thin film of calcium aluminosilicate. Among them were bizarre forms such as ***Opabinia*** (**1**), with a head bearing five compound eyes on stalks, and a long extension carrying grappling hooks to catch prey which could then be conveyed to the mouth.

Some of the soft-bodied fossils from the Burgess Shale can be shown to be related to living animals. The worm-like fossil ***Aysheaia*** (**2**) had a series of body segments, each bearing a pair of fat legs with sharp spines. It is known to have eaten sponges (**3**) – remains of sponge skeletons in the form of tiny spicules of silica are found in its stomach.

Aysheaia is related to the living *Peripatus* which lives under the bark of tropical trees and looks like a velvet coated caterpillar. *Peripatus* is thought to be an intermediate form linking the segmented worms and the jointed-legged arthropods.

How a sponge-feeding, worm-like animal living at the bottom of the sea evolved into an air-breathing predator of tropical forests we do not know, but some

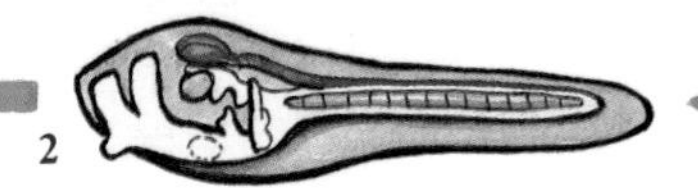

mysteries of the Shale have been resolved. *Peytoia* a circular fossil once thought to be a jellyfish is now known to have been the jaws of a giant arthropod!

The origin of the vertebrates

The first known fragments of vertebrates come from the uppermost Cambrian of Wyoming. They are just scales and pieces of bony armour, but in the following period, the Ordovician (510-438 million years ago) of Australia and North and South America, mud-grubbing, jawless fishes are well known.

It is clear that the ancestral vertebrate was unarmoured, and the Burgess Shale *Pikaia* seems to fit the bill exactly – a free-swimming fish-like animal feeding on minute organisms in the water.

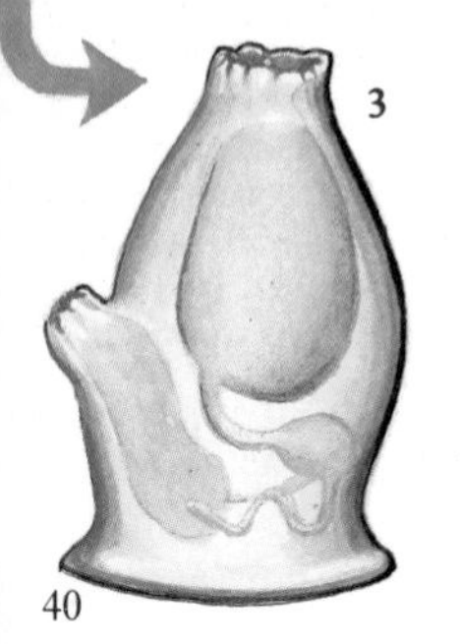

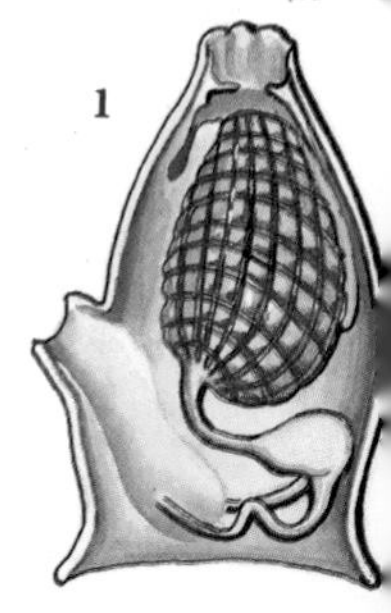

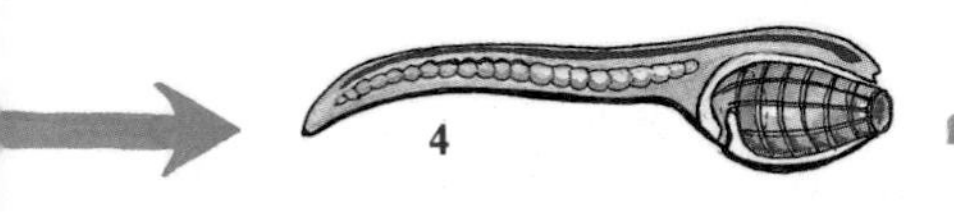

The ancestry of such an animal is suggest by the life cycle of primitive relatives of the vertebrates: the sea-squirts (**1**), which live fixed to the sea bed and filter nutrients from the waters by means of a basket-like structure, the pharynx (*shown red*). These have a larva stage (**2**) which has a swimming tail, a dorsal nerve cord (*purple*), a stiff rod of cells running along the tail (the notochord, *green*) and a perforated pharynx (*red*). All these larval features are the trademark of all vertebrates at some time during their individual life histories.

The larva is the dispersal stage of the animal's life. It is an active swimmer, until it returns to the sea bed to become fixed (**3**). If the larval stage is extended and the larvae develop to sexual maturity (**4**) before returning to the sea bed to become permanently attached adults, a new type of organism will have appeared and the former adult stage eliminated at a stroke. It is believed that such a revolutionary jump produced the first vertebrates. Subsequently, with the formation of a bony armour, the early vertebrates (**5**) returned to life on the sea bed, feeding on the organic debris contained in the muds.

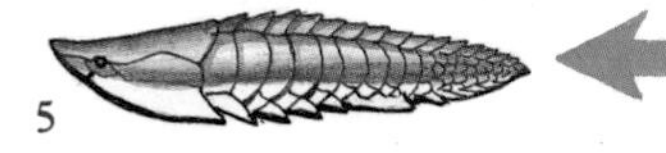

Trilobites

The commonest fossils in Palaeozoic rocks are those of trilobites (**1**) which have an external skeleton divided into three regions: head, thorax and tail. The thorax and tail are composed of several segments, each of which has two pairs of appendages: a set of walking legs and a set of gills. On the head there are well developed compound eyes. In order to grow, the trilobites had to shed their skins periodically and many fossil remains are these cast-offs.

The Cambrian trilobites fed on organic materials on the surface of the sea floor and common Cambrian and Ordovician fossils, known as ***Cruziana*** (**2**) are in fact the feeding trails of trilobites. There are also walking tracks and resting places of trilobites preserves as fossilized impressions.

After the Ordovician (510-438 million years ago) and Silurian (438-410 million years ago) times the trilobites evolved very many different life styles. Some were blind burrowers in the sediments, others developed enormous eyes and spiky outgrowths and probably swam in the surface waters of the oceans. Some, such as *Ampyx*, had forward pointing and elongated projections from the corners of the head shield.

By the time of the Carboniferous (355-300 million years ago) only one family survived and by the end of the Palaeozoic (250 million years ago) all trilobites had vanished and other arthropods, such as crabs, shrimps and their allies, seem to have filled the trilobite niches.

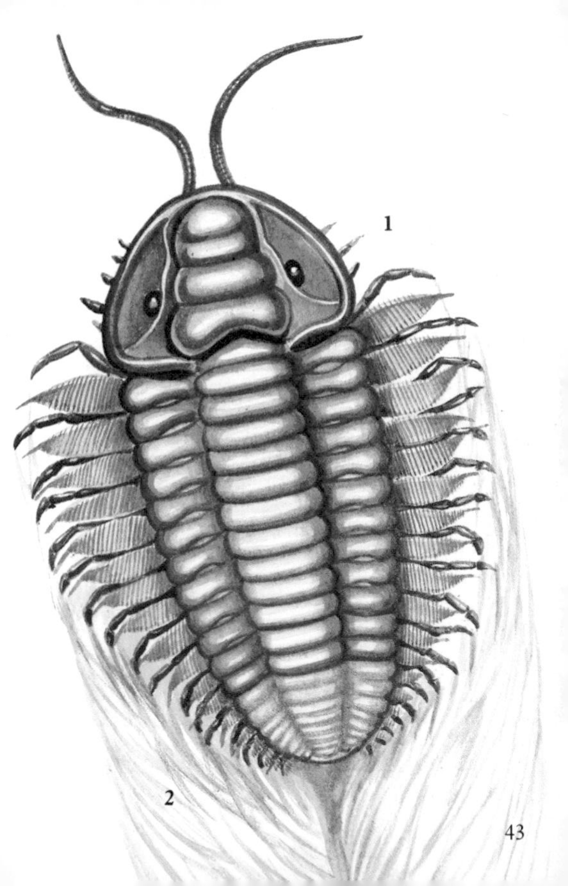
1
2

Molluscs

The molluscs are a major division of the animal kingdom. They all have a muscular foot (*brown*), a head with eyes and tentacles and a calcareous shell (*black*). The most primitive forms come from the early Cambrian. They were thought to have died out in the Devonian until a live specimen was dredged up from the Pacific depths in 1952. This was ***Neopilina*** (**1**). It looked rather like a limpet, except that it had paired gills (*pink*) and paired muscles, proving that molluscs were primitively segmented like worms and arthropods.

Three major types developed. The snails (**2**) used ribbons of rasping teeth to feed off algae and later became carnivores using their teeth to bore holes into the shells of other creatures, or even to inject poisons. The alimentary canal was twisted round, but this is not related to the coiling of snail shells. The bivalves (**3**), became filter-feeders and their head and eyes were lost. Bivalves may burrow into sediment or bore into rocks or wood, or be attached to rocks as oysters and mussels are.

The most advanced molluscs, the cephalopods (**4**), divided their muscular foot into tentacles. Their conical shell contained gas chambers for buoyancy and the early forms laid down heavy calcareous deposits to weight the shells horizontally.

The cephalopods are still among the most successful advanced molluscs and form an important part of the modern food web in the oceans.

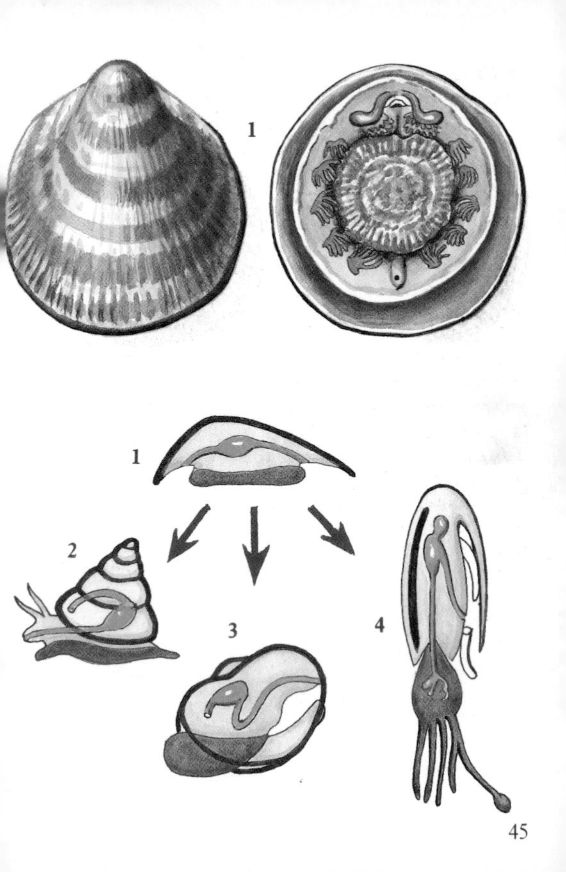
1
1
2
3
4

The first hunters

The cephalopod molluscs were the first real carnivores of the oceans. They developed long conical shells with partitions in which gases collected. By controlling the amount of gas, these animals were able to vary their buoyancy and move up and down in the water. To keep the shells horizontal, heavy deposits of calcium carbonate were laid down as internal weights. Later the animals maintained their stability in the water by coiling the shells.

From the time of the Ordovician period, 510-438 million years ago up to the present day, cephalopods have been among the most successful inhabitants of the oceans, feeding mainly on arthropods, although modern schools of squid hunt fish by driving them up to the surface. Some modern types, such as the octopus, hide in crevasses and lie in wait for their prey. Squids and cuttlefish are fast hunters and their shells are encased inside their bodies.

In Silurian rocks, 438-410 million years old, from Czechoslovakia, primitive cephalopods or nautiloids, have colour patterns preserved. The long cone ***Ormoceras*** (**2**) has straight stripes, whilst the slightly curved cone ***Rizosceras*** (**1**) has zig-zag bands. Although the true colours are not known, the exact patterning is.

In later rocks, cephalopod jaws or rhyncholites (= stone beaks), as well as the tiny hooks from their arms or tentacles are found. Even the ink sacs full of ink are preserved. These rare fossils prove that even these early cephalopods had jaws and tentacles like the living forms.

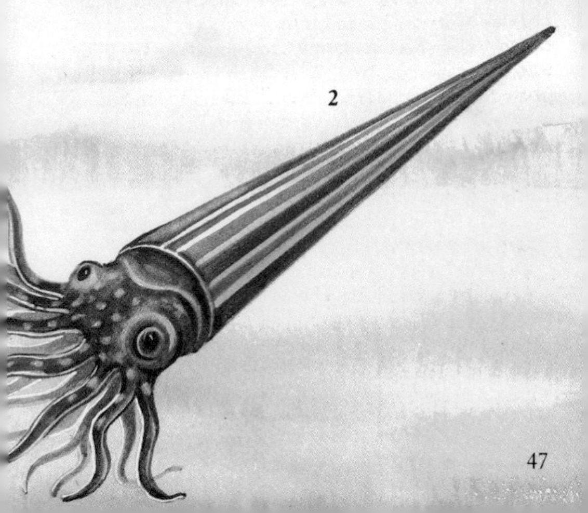

The nautiloids dominated the seas for nearly 100 million years and gave rise to the ammonites. While they hunted trilobites and other arthropods most animals at this time fed on microscopic creatures floating in the water or living in the sands and muds of the sea floor. The trilobites sifted through the sediment, bivalves and other shellfish filtered the water.

Sea-lilies

One group of animals, the echinoderms or spiny skins, which includes starfish and sea urchins, had an internal skeleton of pure calcium carbonate and a five-fold symmetry. There were many more types than the surviving modern forms.

One class of echinoderms, the **crinoids** or **sea lilies** (*illustrated*), were fixed to the floor of the ocean by roots. The long stem which held the body off the sea floor was built up of cylindrical ossicles (= little bones). Surrounding the central body were long, branched food-collecting arms which are in multiples of five. These arms could be fanned out to face into water currents filtering the maximum volume of water to extract small organisms for food.

Another group, the feather stars, has lost its stalk and swims in the surface waters of the oceans.

The crinoids now live only on the floor of the deep oceans but among past fossils were once the most abundant. They make up, for example, the massive Carboniferous Limestone laid down 355 million years ago.

The conquest of the land

At the end of the Silurian period 410 million years ago, the proto-Atlantic or Iapetus ocean closed and a huge mountain chain was formed as North America and Europe collided. This was the time when arthropods, worms and fishes invaded freshwater rivers and streams but by far the most significant event was the emergence of plant life on to land.

Plants, using sunlight to combine carbon dioxide and water, make their own food. Plants at the edge of the water evolved a means of obtaining carbon dioxide from the air by projecting into the air. Two features developed to enable this to happen: a waterproof outer layer, to prevent the plant drying out, and an internal support to hold the plant up in the air. This latter was provided by thickened water-carrying vessels. Hence all such plants are known as vascular (= vessel bearing) plants. The simplest types, ***Rhynia*** (**2**), had smooth stalks with fruiting bodies at the tips. The more advanced ***Asteroxylon*** (**1**) showed the very beginning of leaves, which greatly increased the surface area.

Near Aberdeen, Scotland, 400 million years ago a peat bog was suddenly inundated with hot silica-laden waters from a nearby volcanic eruption, which preserved everything in silica, even the microscopic vessels and the cuticle of the first vascular plants, as well as fungal threads decomposing plant material. Also present were primitive wingless insects, which still survive as silverfish (**3**) and bristletails, as well as tiny mites. These were the first true land animals.

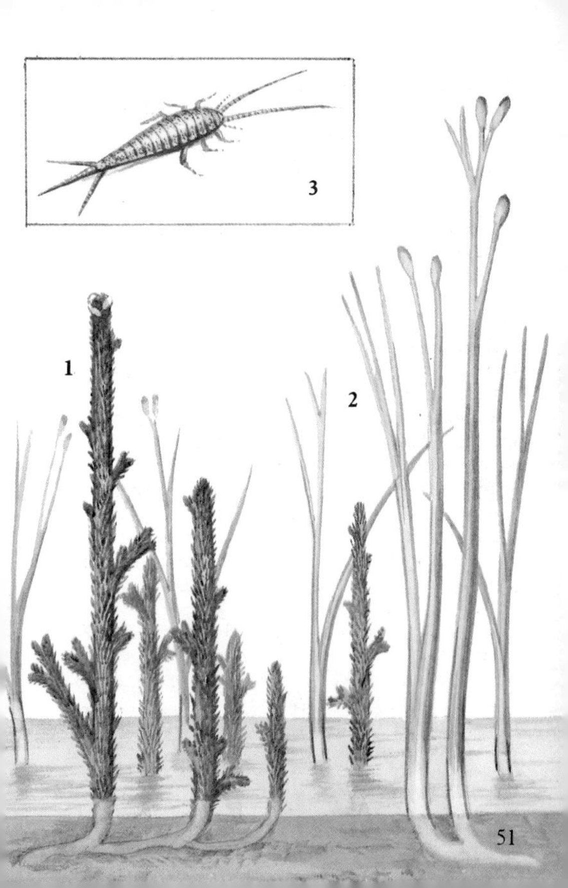
3
1
2

During the Devonian period, from 410 to 355 million years ago, a major change took place in plant life. In the early plants the male and female spores were shed into water where they fertilized and passed through a preliminary water-living stage of development, as mosses, ferns and horsetails still do. In the Devonian some of the large female or megaspores (= large spore) remained attached to the parent plant and were fertilized by tiny wind-blown male microspores. The first plant to achieve this was ***Archaeosperma*** (**1**). In this way the first seeds were formed and from that moment, it was no longer necessary for plants to have an aquatic stage in their life history. From this time plant life could spread over dry land.

Well before the end of Devonian times, 355 million years ago, large tracts of the surface of the land were

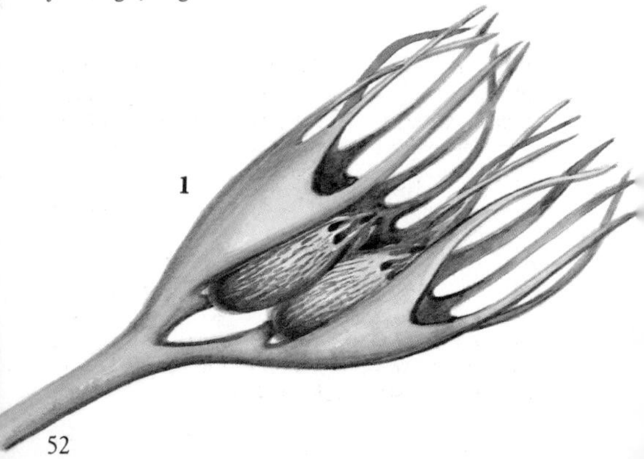

forested. ***Archaeopteris*** (**5**, the ancestor of pines, ***Cyclostigma*** (**4**), a tree-like fern, ***Lepidosigillaria*** (**3**), a giant clubmoss, and the horsetails ***Pseudobornia*** (**2**) all developed. All these were huge, tree-sized plants and formed vast forests.

Once plant life was established on land, animals followed.

The newly evolved plant cover resulted in the formation of soils, a mixture of broken down rock with dead and decaying plant materials. The soils retained moisture and slowed down the rate of erosion, so that there was a fundamental change in the geological erosion cycle on Earth.

Ostracoderms - the first fishes

The first vertebrates, the ostracoderms (= shell skins), were covered with a bony carapace behind which was a scaly tail. They had no jaws or teeth and simply scooped mud from the sea floor up into their mouths to extract nutriments from it. At the end of the Silurian period (410 million years ago) these first fishes invaded the freshwaters of rivers and lakes.

There were three separate and isolated land masses where this happened: the newly formed continent of Euramerica, Siberia and China. In each of these regions, the ostracoderms had their own independent evolution.

The amphiaspids (= shield on both sides) are only known from Siberia. All had a single bony carapace which covered the top and underside of the fish, with a mouth at the front, two orbits for the eyes and a pair of gill openings for breathing. In ***Hibernaspis*** (**3**) the eyes were at the very front with the gill openings also set far forward. In ***Angaraspis*** (**1**) and ***Gabreyaspis*** (**2**) the first gill is converted into a spiracle, a tube which opens through the armour close to the eyes, so that water can be drawn into the gills from above. This is a great advantage in muddy conditions. The outer surface of the bony armour is made of tiny ridges of dentine, which is the main component of teeth. These fishes did not have any teeth, but teeth evolved from the dentine tubercles that formed around the margins of the mouth.

In China a unique group of ostracoderms, the galeaspids (= helmet shield) evolved. These had an

1
2
3

oval opening between the eyes which was first thought to have been the mouth. Later it was realised that it was a single organ for tasting chemicals in the water.

Some of the newly discovered Chinese forms such as ***Duyunaspis*** (*illustrated opposite*) are preserved as lumps of iron ore. The inside of the head region of the galeaspids contained a bony tissue, which surrounded the brain and nerves as well as blood vessels. When the animal died the soft parts rotted away leaving empty holes in the bony armour. Subsequently iron rich waters percolated through the specimen and deposited iron compounds within all the spaces. Later further waters dissolved away the bone to expose a replica of the internal anatomy of the head region. This replication enables the soft parts of the fish to be studied, which in normal fossils is not possible.

In the midline is the simple brain, with cranial nerves, behind is the spinal cord with spinal nerves (*shown yellow*). On either side of the brain is the ear apparatus with two sets of semicircular canals the organs of balance (*green*). There are large veins and sinuses that were filled with venous blood (*blue*) as well as arteries (*red*) coming from the gill region. The narrow paired structures (*pink*) are muscle blocks positioned above the gills. When these Chinese fishes were first discovered it was not known whether they were related to other fishes, but the knowledge of their soft parts proved that they were relatives of the living eel-like lamprey.

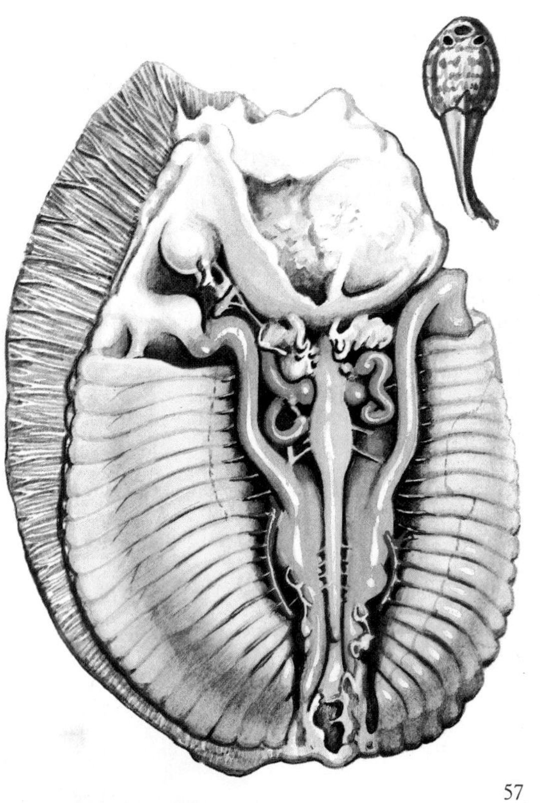

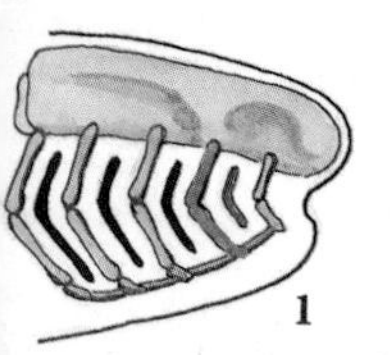

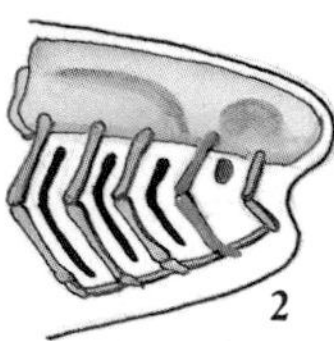

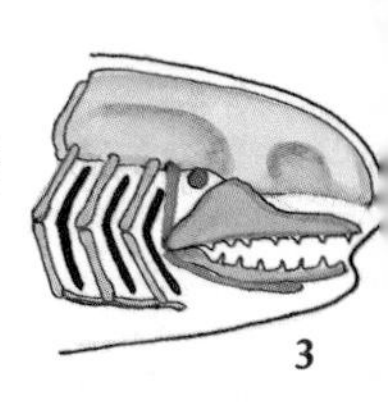

The origin of jaws

One of the greatest breakthroughs in the history of the vertebrates came at the end of the Silurian (410 million years ago). It was the change in function of the front gill supports into jaws. This heralded a new way of life: hunting other animals for food instead of just sucking up nutritious muds.

The primitive arrangement of the gill supports (**1**) shows the first or mandibular (*green*), behind which is the first gill with the hyoidean support (*blue*) behind it. The first step in the origin of jaws was the reduction of the first gill to a spiracle (*red*), just as happened in the Siberian amphiaspids (**2**). The hinged bony gill supports in front of the spiracle then

expanded to fill the space where the gill had been and formed upper and lower jaws (**3**).

The first jawed fish, the acanthodians, or spiny sharks, such as ***Climatius*** (**4**), gave rise to many kinds of bony fish. ***Eusthenopteron*** (**5**) had muscular fins and had lungs as well as gills. During the annual dry season, when the rivers became fouled, these fish were able to breathe air and crawl from one pool to another. These became the first land vertebrates, the amphibians, such as ***Ichthyostega*** (**6**) from Greenland.

Fossil footprints as well as bones of amphibians have been found in Scotland, Russia, Australia and Brazil in rocks of the Devonian period, 300 million years old. This was the beginning of the animal conquest of the land.

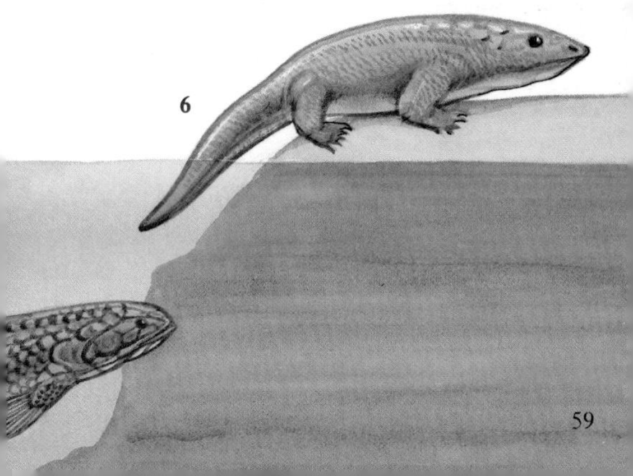

The Age of Amphibians

During the Carboniferous (coal-bearing) period, 355-300 million years ago, there were swamps extending eastwards from North America to Russia, which were all part of the single continent of Euramerica.

Europe was at the Equator and in the climatic tropical high forest zone. The plant life was on a huge scale; there were giant horsetails, giant club mosses, giant ferns, as well as the ancestors of the conifers (*see pages 52-3*). When these plants died and fell into the swamps, they turned into peat and, as millions of years went by, compression and heat gradually transformed the peat into coal.

It was in this environment that the amphibians came into their own. Some became completely adapted to a fully aquatic life, such as ***Crassigyrinus*** (**2**) with limbs reduced and a long swimming tail. Others lost their limbs altogether, so that they looked

like snakes. The main specialization was a flattening of the body so they could move through the shallow waters of the swamps.

There were some amphibians such as ***Eoherpeton*** (**1**) that lived in the drier parts of the forests, feeding on insects and other arthropods. These terrestrial forms developed strong walking legs, and the larger forms preyed upon smaller amphibians. In the swamps the smallest amphibians fed on insect larvae, and fish and the larger swimming amphibians hunted other amphibians.

The coal swamps

When coal is formed by the compression of plant remains many things are preserved: the roots of giant trees, as well as the patterned bark of the trunks, fronds of leaves, fruiting cones and the different types of spore. Often it is not possible to decide which fossils belong to the same type of original plant, so all the different parts have their own names. The giant horsetail stem *Calamites* has leaves known as *Annularia* because for a long time it was not realised that they belonged to each other.

In the swamps there was a profusion of insects and other arthropods. There were harvestmen and the scorpions that preyed on them. There were jumping spiders as well as web-weaving spiders (**5**). There were centipedes (**3**) which also hunted insects, millepedes, including the 2m (6½ft) long *Arthropleura* (*illustrated on page 3*), and cockroaches (**4**) which fed on leaf litter on the forest floor.

The most striking event was the conquest of the air by insects. The giant dragonfly ***Meganeura*** (**1**) hunted other insects on the wing. A primitive stonefly ***Lemmatophora*** (**2**), is remarkable for having three pairs of wings, the extra circular front pair acting as balancing organs.

No vertebrates at this time ate plants. Plant material was broken down by bacteria and fungi, which were themselves eaten by worms and insects, the insects then being eaten by the amphibians, which in their turn were eaten by larger amphibians, which were themselves preyed upon by even larger forms.

1
2
3
4
5
6

Scottish sharks
During the 1980s a major discovery was made of Carboniferous sharks, or cartilaginous fishes, so named because their skeletons were made of gristle or cartilage instead of bone. They were found by Stan Wood on a housing estate in Glasgow, when he was taking his dog for a walk. Among the more bizarre sharks was a 62cm (24in) long complete specimen of ***Stethacanthus*** (**1**). Rising up from its back was a massive fin, which bore what looks like a brush of teeth. Nobody has any idea what purpose this curious structure could have served. Another strange fish

from the same rocks was a 450mm long complete ratfish ***Deltopytychius*** (**2**), which fed on shellfish which it crushed with large flattened teeth.

Together with the sharks there were many kinds of bony fishes, including acanthodians similar to *Climatius* and close relatives of *Eusthenopteron*. There were also numerous spiny shrimps, *Anthracophausia*.

All these animals lived in the sea, although their ancestors inhabited freshwater lakes and rivers. The significance of the vertebrates return to the sea, in particular the sharks, was that they now challenged the cephalopods for the supremacy of the seas. Some fishes fed on shellfish, others on arthropods and from the evidence of preserved stomach contents still others hunted fish.

The first reptiles

Many of the Carboniferous amphibians spent most of their lives on land, but in order to reproduce they had to find open bodies of water to lay their eggs, which then developed into gill-breathing tadpoles. As with most living amphibians this involved laying hundreds or thousands of eggs in the hope that a few would survive to maturity.

There is another approach, which involves laying fewer eggs but providing greater care in the egg itself (**1**). The most effective strategy is to provide a private pond that is not shared by innumerable other individuals. This was accomplished by enclosing the developing embryo in a fluid-filled membrane, the amnion (*blue*). A food supply in the form of a yolk (*yellow*) and a respiratory membrane, the allantois (*purple*), to enable the developing embryo to breathe,

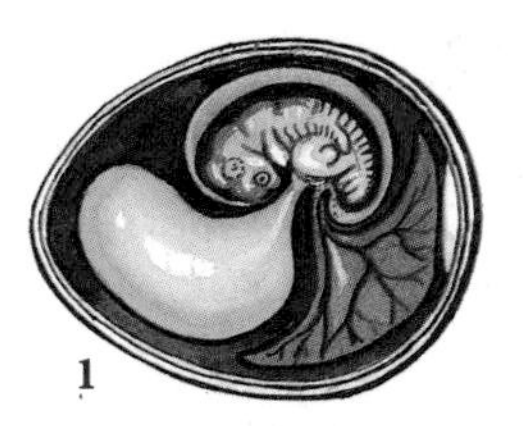

were enclosed with the amnion in a protective membrane, the chorion (*black*), around which a porous shell of calcium carbonate was deposited. The animals with this amniote or cleidoic (= closed) egg, are called reptiles.

The first reptiles, such as ***Petrolacosaurus*** (**2**), lived in the Carboniferous swamps and hunted insects. These small lizard-like reptiles were first discovered in Nova Scotia, Canada, preserved in hollow tree stumps into which the animals must have fallen and been unable to get out again.

Primitive paramammals

With the advent of the Permian period, 300 million years ago, the tropical swamps dried out. The reptiles that could lay their eggs on land suddenly had an enormous advantage: they were not obliged to return to the water to breed.

The dominant life on land was the paramammals, or mammal-like reptiles, which were the precursors of the mammals. The smallest fed on insects and worms, the larger ones on the smallest and they in their turn were preyed upon by the largest. All the reptiles fed on other animals, none had learnt how to eat vegetable matter directly. The continents were ruled exclusively by flesh-eaters.

One of the problems faced by large reptiles was how to control their temperature and the large ***Dimetrodon*** *(illustrated)* developed a large vertical sail along its back supported by extensions of the vertebrae. When the animals faced into the sun it acted as a cooling membrane and when the main area was faced towards the sun's rays the animals would warm up. This meant that with this device, these reptiles could be off hunting earlier in the day than forms without sails.

During the middle part of the Permian period a few plant-eating forms began to evolve in Russia and these gradually became the commonest of all the paramammals. This was a major step in the evolution of food chains. A large proportion of plant-eaters was preyed upon by a small minority of flesh-eaters – a modern type community structure had developed.

Advanced paramammals

During the Permian there were changes in the paramammals; many began to walk upright instead of sprawling like typical reptiles. The teeth became different; there were front teeth for tearing and stabbing and back teeth for chewing. What was especially important was that some seem to have developed a furry covering - which could bring the ability to control their internal temperature.

Perhaps the most striking advance was the evolution of the plant-eating dicynodonts (= two dog teeth), which were toothless apart from a pair of tusks in the upper jaw. They were of many shapes and sizes, including the squat hippopotamus-like ***Lystrosaurus*** (**1**). There were vast numbers of these and only a small proportion of flesh-eaters, such as the dog-like ***Cynognathus*** (= dog jaw, **2**) and the sabre-tooth forms found in Russia.

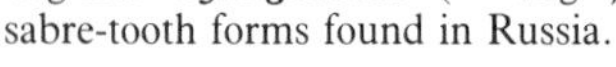

It was at this time that the first modern style food web evolved on land, with a large number of plant-eaters being preyed upon by a small number of carnivores. Towards the end of the Permian many paramammals had become very like mammals and it looked as if the age of mammals was about to dawn. But then, after having dominated the continents for 70 million years, they suddenly went into decline.

A minor feature of the life of the times were small insect-eating lizard-like reptiles but they gave rise to modern reptiles and dinosaurs.

During the Permian, the Uralian Sea closed and Europe and Asia became joined together. This was the final continental collision that united all the land of Earth in the single continent of Pangea.

The Great Permian Death

At the end of the Permian period, 250 million years ago, there took place the greatest setback to existing species ever experienced by life on this planet. It resulted in the greatest mass extinction of all time, more than 95% of all species on Earth died out.

When this extinction is examined in detail it can be seen that marine animals had been dying off gradually over the previous 30 million years. This had happened because the seas were changing and retreating. World-wide there was a lowering of sea level and this in turn caused the chemical nature of the seas to slowly alter. Oxygen-depleted regions were created that spread and wiped out most of the life on the floor of the continental shelves.

As a result of these changes some major groups, such as the trilobites, completely died out during the Permian. However, viewed overall, there were relatively few of the major groups that became completely extinct with all the types of animal in them disappearing.

What was more dramatic in effect was the decimation of types within those groups that did manage to survive, for example of one major shellfish phylum over 125 genera were reduced to only two. Of the 16 families of advanced coiled cephalopods (ammonites) only one survived.

The abundant life in shallow-shelf seas was virtually eliminated. Life on land and in the ocean depths does not seem to have been subjected to this terrible crisis.

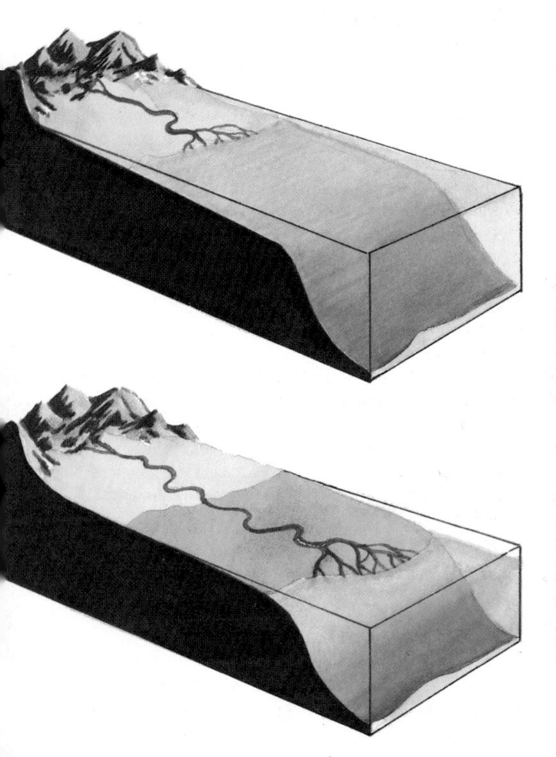

The Mesozoic Era: The Triassic Period
(250-205 million years ago)

At the beginning of the Triassic period, 235 million years ago, there was a rise in sea level and marine life quickly returned to its former diversity with a proliferation of all kinds of shellfish.

On land the most important changes in vertebrate life took place. At the beginning paramammals dominated, with vast herds of plant-eating dicynodonts, such as ***Dinodontosaurus*** (**1**). The main carnivores were the furry cycodonts. There was, however, a group of flesh-eaters, the thecodonts (= teeth in sockets) including ***Rauisuchus*** (**2**), that

returned to the water and were crocodile-like in their way of life. These were the ancestors of the dinosaurs.

For some unknown reason the dicynodonts began to go into a rapid decline. They were replaced by heavily built lizards with a strong hooked beak and a battery of crushing teeth, the rhynchosaurs. They included the plant-eating **Scaphonyx** (**3**). The carnivorous cynodonts also began to die out but they gave rise to a group of water vole-like forms as well as tiny insect-eating true mammals. The thecodonts came on to land and ousted the cynodonts. By the middle of the Triassic period the age of paramammals was over. There were a few surviving giant-sized dicynodonts but the landscape was dominated by rhynchosaurs.

The Age of Dinosaurs

Towards the end of the Triassic, 195 million years ago, there was a second major change in life on land. The huge numbers of rhynchosaurs that had flourished throughout the world suddenly vanished, the last herbivorous dicynodonts became extinct, even the carnivorous cynodonts disappeared. One small group of herbivorous parammamals survived by occupying the ecological niche of the living water vole: ***Oligokyphus*** (**4**). The cynodonts gave rise to a number of tiny, furry, shrew-like true mammals. These became the main animals of the night, hunting insects, snails and worms, once night fell.

The dominant reptiles were the descendants of the thecodonts, the dinosaurs. The flesh-eaters, the theropods (= beast-footed), walked on their hind feet; there were two groups, the heavily built large-headed carnosaurs (= flesh reptiles) and the lightly built coelurosaurs (= hollow reptiles) with long necks and small heads. Among the coelurosaurs was ***Coelophysis*** (**2**).

The coelurosaurs gave rise to the four-footed plant-eating prosauropods (= first lizard-footed), such as ***Massospondylus*** (**3**), replacing the rhynchosaurs and large dicynodonts, which then gave rise to the giant sauropods, the brontosaurs. A group of small two-legged, plant-eating thecodont descendants, the ornithopods (= bird-footed) such as ***Lesothosaurus*** (**1**), took the role of the small herbivorous paramammals. The age of dinosaurs had dawned.

1
2
3
4

The origin of the dinosaurs

The event that overshadowed everything else towards the end of the Triassic was the appearance of the dinosaurs.

The primitive thecodonts were rather like crocodiles, with bony plates under the skin and a heavy tail flattened from side to side, for swimming. The limbs stuck out sideways in the typical sprawling manner of reptiles (**1**). The advanced thecodonts had their legs partly straightened so that the body was held off the ground exactly as in the living crocodile's high walk (**2**). The hind limbs became longer and stronger as they produced the initial thrust to drive them through the water. They could obtain a more powerful movement by swinging the entire limb from the hip and shoulder joints.

When such reptiles came out of water, the stride, and hence speed, of their hind legs was much greater than the front, so the front part was lifted off the ground and the heavy tail acted as a balance to the trunk and head. These were the dinosaurs, belonging to two separate groups, the saurischians (= lizard-hipped, **3a**) and ornithischians (= bird-hipped, **3b**). Fundamentally they are both reptiles that can walk and run in the same way as mammals, with a fully upright posture. The upper bone of the pelvis, the ilium (*cream*) is fused to the backbone; the pubis (*purple*) faces forwards but in the ornithischians it swings backwards and a new forward projection, the prepubis, is formed; the ischium (*blue*) always points backwards.

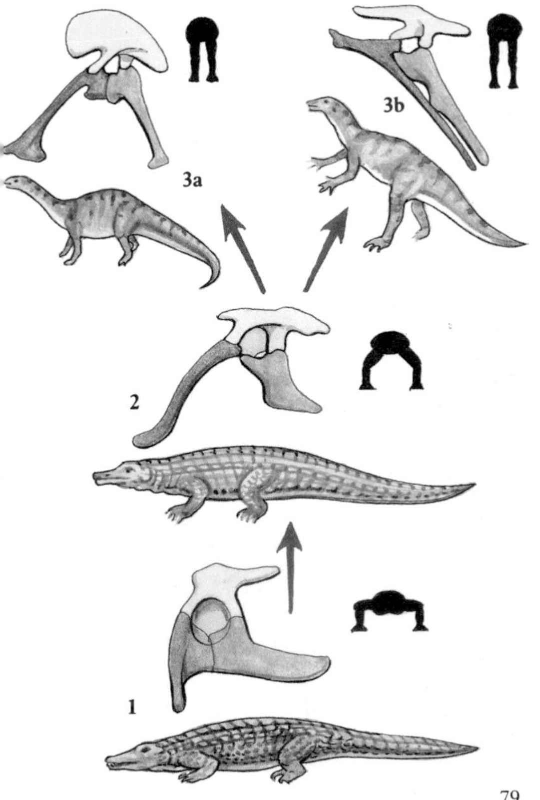
3a
3b
2
1

The first dinosaurs

The appearance on land of two-legged, flesh-eating dinosaurs was a fairly sudden event. Such speedy killers were so efficient that the ponderous plant-eaters, the dicynodonts and the rhynchosaurs, stood little chance against them. The most primitive were the heavily built carnosaurs. They had large skulls, short thick necks and bore a strong armour of bony scutes just beneath the skin. They were distinguished by the structure of the hind limbs and the hip girdle. One of the best known is ***Ornithosuchus*** (**1**) from the Triassic of Scotland.

A second group were the coelurosaurs (= bird-crocodile), which were much more graceful looking. They had long legs and a small head on a long neck. Their teeth, like carnosaur teeth, had serrated edges

and were flattened and blade-like. There is no doubt that these were active flesh-eaters and some skeletons of ***Coelophysis*** (**2**) from America have the remains of young individuals in their stomachs, suggesting that they must sometimes have eaten their own young.

Both kinds of flesh-eater were very common and many Triassic sands are covered with their tracks showing that they were walking at about 8 kph (5mph). When first discovered in Connecticut they were described as Noah's ravens because their three-toed prints were so bird-like.

Plant-eating dinosaurs

Plant-eating among the dinosaurs began with a branch of the coelurosaurs, which gave rise to the prosauropods (= first reptile-footed). They still had

long necks and small heads, with numerous sharp teeth, but their legs were proportionately shorter and their bodies more heavily built. Their overall proportions reveal their coelurosaur ancestry but the prosauropods marked a fundamental change in the dinosaur lifestyle. In spite of still having flesh-eating teeth they had developed the ability to digest plant materials.

With their increase in size, these were also the first dinosaurs to become 'warm-blooded': they maintained a constant internal temperature by virtue of their increased volume. This was because, in relation to their volume, their surface area was relatively too small for them to be able to heat up or cool down rapidly.

A small form 2m (6½ft) long *Thecodontosaurus* was discovered in the middle of the city of Bristol in the

remains of a collapsed cave deposit. One from Argentina, the mouse dinosaur *Mussaurus*, was only 20cms (6½in) long. With this size it is difficult to be certain that they were no longer flesh-eaters but had become herbivores. However, the size of the main types such as ***Plateosaurus*** (*illustrated*) from Germany, which was 6m (20ft) long, and the even larger 12m (40ft) long *Riojasaurus* from the Argentine, and the fact that from the evidence of footprints, they lived in large herds, show that they must have been plant-eaters. Furthermore, they gave rise to the gigantic herbivorous brontosaurs (= thunder lizards). At the end of the Triassic a group of small 1m (40in) long bipedal plant-eating ornithischian dinosaurs such as *Lesothosaurus* appeared, some of which slept throughout the hot summers in burrows, just as some crocodiles do in Africa today.

The return to the sea

With the rise in sea level and the rapid recolonisation of the seas by the surviving shellfish, there was an enormous increase in fishes and cephalopods, especially the ammonites (= ram's horn) in the oceans. This abundant food source lead to several groups of reptiles returning to the sea, which heralded a major change in the economy of the oceans. From now on air breathing vertebrates were the top predators, feeding on fish and cephalopods, the previous top animals of the seas.

By the dawn of the Triassic all the ancient continents were united in the single supercontinent of Pangea, the northern arm of Laurasia and the southern Gondwana with the ocean of Tethys between. Living along the northern shores from England across Europe to China and the southern from Tunisia to Israel and India were nothosaurs (= southern lizards) like ***Nothosaurus*** (*illustrated*). Fossil footprints show that nothosaurs had webbed feet; the bones of the tail suggest that there was a long dorsal fin for swimming. The jaws were long and armed with numerous pointed teeth which acted as fish traps. Remains of dozens of young ones are found in fossil sea caves in what is now southern Poland.

Shellfish-eating reptiles

During the middle of the Triassic period there appeared a unique group of reptiles, the placodonts (= flat-plated teeth). These had a thick bony carapace reminiscent of turtles (though not related to them) but with a long tail with a dorsal fin. The feet were webbed. Some forms such as *Placodus* had peg-like teeth at the front of the jaws, whilst ***Placochelys*** (= plated turtle, *illustrated*) had a toothless beak; these structures were for pulling up shellfish that were fixed to the sea bed. The roof and floor of the placodont mouth had a pavement of large flat teeth which served to crush the calcareous shells of molluscs and other shellfish.

Placodonts were confined to shallow tropical seas which covered Germany, as well as Tunisia and Israel. The sudden increase in the amount of shellfish with the rise of sea level in the Triassic produced a food resource that was unexploited, for the placodonts remained restricted to this one small region and did not spread to other parts of the world. They could only survive in very coastal regions. They filled this niche for a time, but they were unable to compete with the bony fish that developed powerful jaws and shell-crushing teeth. Unlike air-breathing reptiles, fishes can stay at the sea bed, and so the placodonts died out by the end of the Triassic.

At the end of the age of dinosaurs, 70 million years ago, a group of marine lizards became shellfish crushers. Today walruses fill this same niche.

Gliding and flying reptiles

At the end of the Triassic the vertebrates took to the air.

One of the most amazing reptiles was discovered in Kirkizstan, in Soviet Central Asia, and is named ***Sharovipteryx*** (= Sharov's wing, *illustrated*) after its discoverer G. Sharov of Moscow. It was originally named *Podopteryx* (= footwing) but this name had already been used by someone else for a quite different animal. The skeleton looks just like any ordinary lizard-like reptile but its superb preservation shows that there was a membrane of skin joining the elbows and knees. There was also an equally extensive membrane extending from the ankles to part-way down the tail. When this reptile leapt out of the trees it would have stretched its limbs out to make a perfect glider and could have planed over long distances in exactly the same way as the living flying squirrel.

It seems likely that from just such a glider the true flying reptiles or pterosaurs originated. The membrane attached to the trailing edge of the forelimb, at the elbow, if extended would attach to the last finger. Elongation of that finger would have enlarged the membrane into a triangular wing, which would still have been attached to the hind legs. This is the basic pattern of the pterosaur wing as seen in the first known example *Eudimorphodon* from the Triassic of Italy. The pterosaurs developed true flapping flight and the group dominated the skies for 140 million years.

Parachutists

Many different ways of taking to the air seem to have evolved. In Soviet Kirgizstan, Dr Sharov discovered a fossil reptile he named ***Longisquama*** (= long scale, **2**). It had rows of enormously elongated scales running outwards along the length of its back, which must have acted as a kind of parachute when it leapt out of the trees. Originally it was believed that there was a single row of these scales, but it has now been shown that they were paired. The scales on the trailing edge of the forelimbs were also elongated to about seven times that of normal scales. This condition may have been the very beginning of the evolution of feathers, which were derived from elongated reptile scales. *Longisquama* was too highly specialised to have been ancestral to birds but it shows how feathers could have originated.

At three separate geological periods, the Permian, the Triassic and the present day, reptiles developed membranes supported on elongated ribs, which could be extended for gliding. In the Permian of England, Germany and Madagascar *Weigeltisaurus* bore a gliding membrane supported by long thin ribs, which were very similar to those of the living flying dragon *Draco* of the tropical forests of Asia. At the end of the Triassic in England the gliding lizard *Kuehneosaurus* and the North American ***Icarosaurus*** (**1**) developed long deep ribs so that the membrane covering them formed the aerofoil shape of an aeroplane's wing. These deep ribs were jointed and could be folded back along the side of the body.

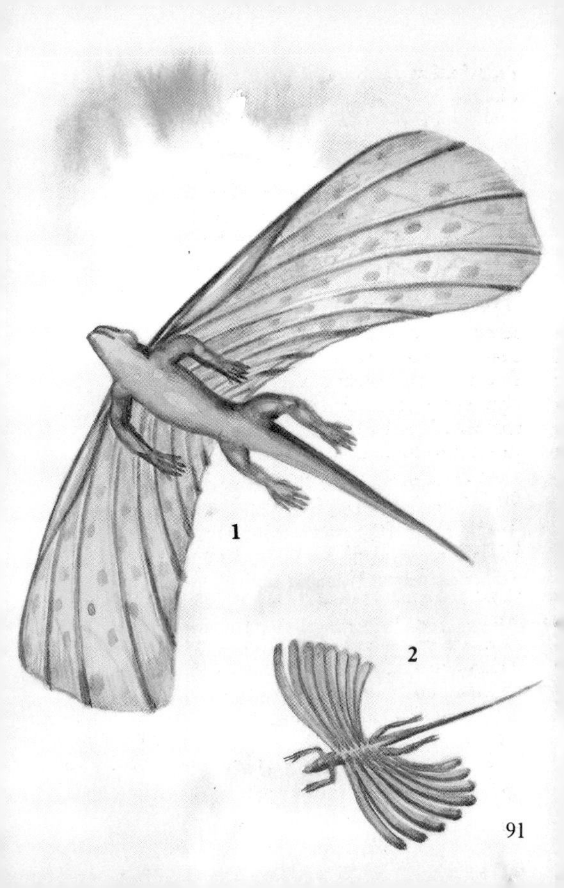
1
2

The Mesozoic Era: The Jurassic Period
(205-135 million years ago)

Ammonites

During the Jurassic period, 205 to 135 million years ago, the supercontinent of Pangea began to break up. The northern and southern parts separated. Antarctica and Australia split away from South America and Africa. The sea flooded between Europe and Asia. This spread of the sea led to a great flourishing of marine life. Vast numbers of shelled cephalopods, the **ammonites** (*illustrated*), evolved from the coiled nautiloids (*page 47*) and hunted in the world's oceans. The empty ammonite shells covered huge areas of the sea floor, forming layers of rock made up of thousands of packed ammonite shells and providing a firm support for shellfish such as oysters.

The ammonites had coiled shells with a series of gas-filled chambers which made them buoyant and also ensured that the last body chamber was always positioned below the main part of the shell. They used tentacles for capturing their prey and swam by jet propulsion, squirting water out through a fleshy tube, the siphon. The males and females had different types of shell, but it is not known which type belongs to which sex. The ammonites were the most successful invertebrate animals in the sea until the very end of the age of dinosaurs, 65 million years ago. They evolved very rapidly and the different types of ammonite are used for dating rocks.

The fish lizards

In 1811 a ten year old girl, called Mary Anning, excavated a large reptile skeleton from the Jurassic rocks of Lyme Regis on the south coast of England. This was the first recognisable specimen of an ***Ichthyosaurus*** (*illustrated*) or 'fish-lizard', although vertebrae had been described in 1712 by John Morton. The ichthyosaurs are one of the classic examples of convergent evolution. Although a reptile it has the general proportions of the dolphins which live a similar kind of life.

All the complete skeletons had a downturn in the bones of the tail and this was interpreted as meaning that there was a compensatory dorsal tail fin. Subsequently, in southern Germany, complete skeletons with outlines of the skin preserved proved this to be the case and also suggested that there was a large triangular dorsal fin as in dolphins and sharks.

The ichthyosaurs were unable to get on to land to lay their eggs and it is known that they gave birth to live young. A few specimens are known where the mother died at the very moment that the young were being born. In 1987 an ichthyosaur was discovered in Somerset with a minute embryo preserved coiled round in a ball, suggesting that it was held within egg membranes, which means that they produced eggs which hatched within the mother.

The main food of ichthyosaurs was cephalopods because even in young 1.5m (59in) long individuals up to half a million chitinous hooks from cephalopod tentacles are found in their stomachs.

Plesiosaurs

Another new type of reptile discovered by Mary Anning at Lyme Regis was the plesiosaur, which was described in 1822 as being like a serpent threaded through a turtle. These reptiles had a barrel-shaped

body, short tail, long neck and small head with numerous needle sharp teeth and four paddle-like limbs.

Another type was soon discovered that had a large head and short neck and with much more stubby teeth. These were named pliosaurs and included the giant ***Liopleurodon*** (*illustrated*).

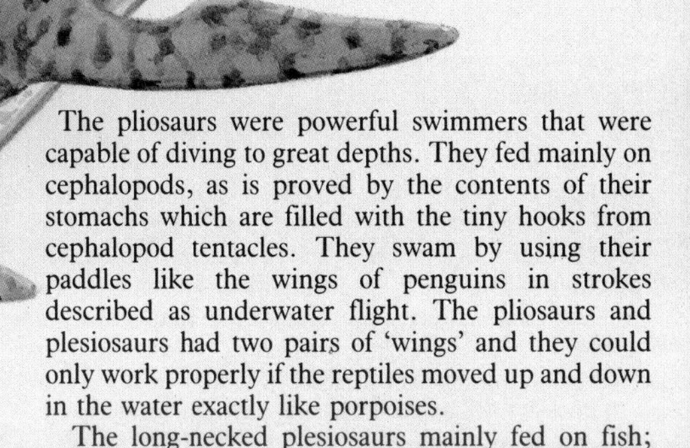

The pliosaurs were powerful swimmers that were capable of diving to great depths. They fed mainly on cephalopods, as is proved by the contents of their stomachs which are filled with the tiny hooks from cephalopod tentacles. They swam by using their paddles like the wings of penguins in strokes described as underwater flight. The pliosaurs and plesiosaurs had two pairs of 'wings' and they could only work properly if the reptiles moved up and down in the water exactly like porpoises.

The long-necked plesiosaurs mainly fed on fish; they were unable to dive and spent most of their time paddling at the surface swinging their heads and long necks over the surface of the water to plop down on to fish. The structure of the bones of the paddles shows that the long-necked plesiosaurs were able to move their paddles very rapidly and were hence highly manoeuvrable; they could twist and turn very rapidly but they were not powerful swimmers.

Crocodiles

At the end of the Triassic period there were already reptiles that looked like crocodiles and filled the ecological niche of crocodiles – these were the phytosaurs (= plant lizards) and they had their nostrils far back from the tip of the snout situated just in front of the eyes. True crocodiles have their nostrils at the tip of their snouts and there is a long bony palate separating the air passages from the cavity of the mouth.

The first true crocodiles, such as the Triassic ***Terrestrisuchus*** (**1**) from the west of England, were lightly built, long-legged, land-living reptiles and it was not until the Jurassic after the extinction of the phytosaurs that they returned to water and developed their long squat bodies with short legs. They filled the empty niche of semi-aquatic flesh-eater which the thecodonts had occupied in the Triassic. During the Jurassic some crocodiles became fully adapted to life in the oceans, their limbs became paddle-like and their tails developed a dorsal fin.

Towards the end of the age of dinosaurs some developed into giants nearly 20m (65ft) in length . These became extinct at the end of the age of dinosaurs but the more usual-sized crocodiles such as ***Sokotosuchus*** (**2**) from Nigeria simply continued well after the end of the age of dinosaurs. Although crocodiles seem to be typical primitive reptiles all their specialisations for a semi-aquatic life are later developments which give a false impression of primitiveness.

1
2

Jurassic land plants

The plants that dominated the surface of the continents during the early part of the age of dinosaurs were the gymnosperms (= naked seeds, the seeds are not enclosed as they are in the fruits of flowering plants). There are two main groups the cycads and their allies and the conifers and their relatives.

The cycads have a massive trunk with a crown of leathery evergreen leaves and they look just like modern-day palm trees, although they are not at all related to them. One of the most common Jurassic forms is **Williamsonia** (**2**) which looks like a palm tree with branches. This type, however, unlike true cycads, had flower-like reproductive parts. ***Cycadeoides*** (= cycad-like, **1**) looks almost identical to living cycads but also has flower-like reproductive parts. ***Palaeocycas*** (**3**) is a true cycad but is much more like a palm tree in appearance.

The conifers or cone-bearing trees contain three basic groups. The true conifers in which the reproductive organs are borne on separate cones, the female ones being large and woody, are the typical pine cones. Yews are different in that there are no female cones but the seed is surrounded by a brightly coloured fleshy cup, the arimsaril. The third type includes the maidenhair trees (*pages 122-3*). Conifers formed a major part of the diet of herbivorous dinosaurs.

The largest dinosaurs

The largest land animals of all time were the plant eating sauropods (= lizard feet), the descendants of the prosauropods. The largest such as ***Brachiosaurus*** (= arm lizard, *illustrated*) weighing up to 100 tonnes. The long-tailed *Diplodocus* 23m (75ft)long was a mere 10 tonnes, *Apatosaurus* (the original *Brontosaurus*) 30 tonnes. The Chinese *Mamenchisaurus* (*illustrated pages 4-5*) had the longest of all dinosaur necks.

All these sauropods were of enormous size and were important plant-eaters. The typical sauropods such as the brontosaurs (= thunder lizards) had teeth at the front of the skull which formed a kind of rake for collecting up plants. The nostrils were on the top of the head. The massive columnar limbs had one claw on the front legs and four blunt hooves or nails, while on the hind feet there were three claws and two hooves.

The second group, the brachiosaurs, unlike all other dinosaurs, had longer front legs than hind ones. There were peglike teeth all round the jaw not just at the front. The nostrils were situated on a rounded projection on the very top of the head.

From the evidence of fossil footprints the sauropods lived in herds of up to 30 individuals, the younger ones walked in the middle with the largest ones on the outside.

The sauropods can be aged by counting growth rings in the limb bones, exactly like the annual rings in sections of tree trunks, and some individuals lived for 120 years.

Swimming brontosaurs

The bones of the limbs of the sauropods are enormously heavy and strong, while the vertebrae of the backbone are light and much of the bone has been excavated so that bone is only formed along lines of force. It has been suggested that the limbs were like diver's heavy-weighted boots to keep brontosaurs down in the water and the light backbone was to help the upper part of the body to float.

In Texas there are footprint tracks of the brontosaur ***Apatosaurus*** (*illustrated*) which comprise only prints of the front feet. Brontosaurs could not do handstands; the only explanation is that these prints

were made by a brontosaur swimming and just pawing the bottom with its front feet much like hippopotamuses do in rivers and lakes today. When they wanted to change direction one of the back feet would be brought down so that the animal could swing round and then paddle off in a new direction. In fact where this Texan trackway changes direction there is a single print of a hind foot.

It was long believed that the brontosaurs could always escape from the flesh-eating dinosaurs by taking to the water. However, in the last few years some further trackways have been discovered that were made by large swimming carnivores: the tips of their toes just touching the bottom.

Flesh-eating dinosaurs

During the Jurassic the two groups of flesh-eating dinosaurs, the large carnosaurs (= flesh lizards) and small coelurosaurs (= hollow lizards) evolved in opposite directions: the more heavily built carnosaurs grew progressively larger till they reached about 12m (39ft) in length, while some of the lightly built types, such as ***Compsognathus*** (**2**), ended up much the same size as domestic chickens about 60cm (2ft) long (remember that chickens do not have long tails).

The giant flesh-eaters, such as the 12m (39ft) long American *Allosaurus* and the somewhat smaller

English ***Megalosaurus*** (= giant lizard, **1**), which was the first dinosaur discovered in 1819, were most likely not active hunters but rather scavengers on dead and dying dinosaurs. Evidence pointing in this direction is found in America with the remains of a brontosaur's tail with scratch marks made by the teeth of carnosaurs, together with some broken teeth. It is clear that a carnosaur was eating bits of meat off a brontosaur tail – a hunter would have been concerned with ripping open a dinosaur and wolfing down the entrails, not picking at what was left of the tail.

The miniaturized coelurosaurs were, in contrast, active hunters and the evidence for this is seen in the remains of lizards preserved in their stomachs.

From the evidence of trackways the smaller types hunted in packs, while the giant ones tended to wander about in ones and twos.

Spiky dinosaurs

Survival of herbivores against the attentions of the flesh-eaters was solved in a variety of ways. Perhaps the most striking type of defence was invented by the stegosaurs, the so-called roofed reptiles, because they developed huge bony plates that were first thought to have formed a kind of roofing.

The ancestor of the spiky dinosaurs is *Scelidosaurus* from the Jurassic of Lyme Regis. This was an armoured herbivore with two rows of small triangular bony plates running the length of the body from just behind the head to down along the tail.

A primitive stegosaur, *Tuojiangosaurus* from China, has a pair of sharp spikes near the end of the tail which were the main defensive weapons and running along its back were small triangular plates. From this basic type there were two contrasting developments.

One is shown by the African ***Kentrosaurus*** (*illustrated*) where from the middle of the back towards the tail the plates are developed into long sharp spikes with a pair sticking out sideways over the hip region.

The other development is seen in *Stegosaurus* itself with huge flat bony plates running over the back. These stood upright and had a large blood supply. They acted as a kind of cooling system to get rid of excess heat.

Herbivore herds

The dinosaurs that became the dominant plant eaters were the ornithischians. They were mainly bipedal (two-footed) but they could walk on all fours and, normally, they fed in this way. Only when they were on the move did they get up on their hind limbs.

The basic stock were the ornithopods (= bird foot). The small forms, such as the 1m (39in) long ***Hypsilophodon*** (= high-ridged tooth, *illustrated*) from the Isle of Wight, were fast runners, although at first it was wrongly believed that they could climb trees. The large *Camptosaurus* was 3m or 4m (10-13ft) long and was much slower. There were two major types: those that remained small and unspecialised and those that became large, weighing several tonnes.

It is not known how these very successful dinosaurs managed to protect themselves but it is known that they lived in herds.

These ornithischians were efficient plant-eaters. They had rows of teeth for shearing up tough plant material, which they plucked with their horny beaks at the tip of their jaws. They were able to grind up plant material in their mouth because the air and food passages were separated with a secondary palate, as found today in mammals. Furthermore they had muscular cheeks, again as in mammals, but not found in any other kind of reptiles.

Furry flying reptiles

Throughout most of the age of dinosaurs the skies were dominated by the flying reptiles or pterosaurs. In 1971 a complete skeleton was discovered in Jurassic rocks in Soviet Central Asia in Kazakhstan in which the wing membrane was clearly preserved but much more remarkably the covering of the body and limbs was also preserved and this proved that the pterosaurs were furry. This pterosaur was named ***Sordes pilosus*** (= hairy devil, *illustrated*) and is one of the key pieces of evidence that show that the pterosaurs were warm blooded in the same way as birds and mammals. Because of this they should not really be classified as reptiles any longer but placed in a major class of their own, the Pterosauria.

The wing membrane of pterosaurs was attached to their hind limbs and in this way the shape of the membrane could be accurately controlled by the hind legs during active flapping flight. The primitive pterosaurs had long bony tails and teeth, which were lost in the later forms.

Recently it was suggested that pterosaurs walked on their hind legs like small dinosaurs but the structure of the hip girdle showed that they had a very awkward sprawling waddling gait which was, however, ideal for clambering up cliffs or tree trunks. Recent discoveries in the petrified forest of the Kyzylkum Desert in Kazakhastan suggest that the pterosaurs did nest in trees.

Furry mammals

Fossil lower jaws of tiny rat-sized mammals were found in Jurassic rocks near Oxford in 1764 but their importance was not recognised until they were brought to William Buckland in Oxford in the 1820s. When he announced their discovery in 1824 it caused a sensation because it was believed that no mammals existed at that geological period.

The fossil record of mammals for the entire age of dinosaurs comprises numerous isolated teeth and a few skulls especially from the Triassic of China. In the 1970s a complete skeleton of a Mesozoic mammal was discovered in Jurassic rocks of Portugal. This mammal skeleton had a hip girdle similar to that of the living marsupial pouched mammals and a long tail that suggested it was used to help it balance in the trees.

The fundamental distinguishing feature of the mammalian skeleton is in the ear region, where the original bones of the reptilian jaw joint (*see colour key*) have become part of a string of sound conducting bones. There were many other features of the tiny mammals that were important. From tiny pits on the bones of their snouts, it is clear they had whiskers and from this it is inferred that they were furry. The teeth had become specialised into pulling incisors, stabbing canine teeth and chewing molars with sharp cusps. The foot and air passages were separated so that they could chew their food and breathe at the same time. This also meant that young could suckle from their mothers.

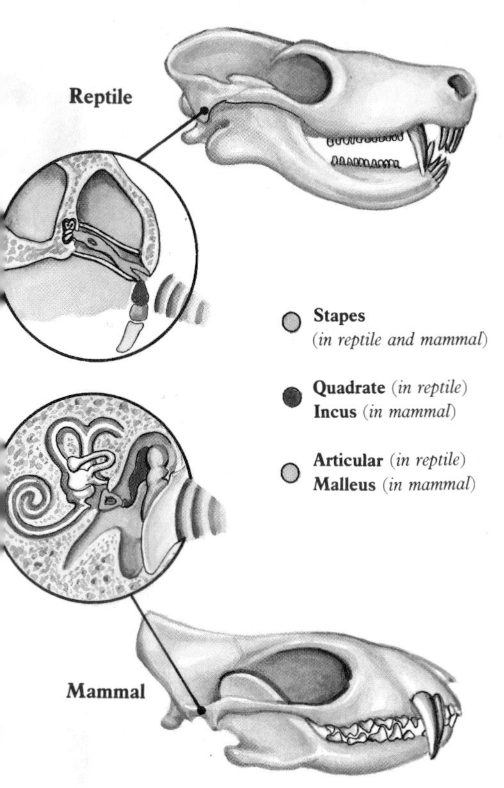
Reptile
Stapes
(in reptile and mammal)
Quadrate (in reptile)
Incus (in mammal)
Articular (in reptile)
Malleus (in mammal)
Mammal

Jurassic lagoon at Solenhofen

In southern Germany there is a fine-grained lithographic limestone of late Jurassic age, about 145 million years ago. It was once mined for use in printing but the owners of the quarries discovered in the last century that high prices would be paid by collectors for the fossils that were uncovered. This limestone was deposited as a very fine limey mud in a lagoon between land to the north and coral reefs to the south. The limestone was formed just below low water mark. This is because all the fossil tracks are only of marine animals, which means that the muds were never directly exposed to the air. The preservation is excellent and in many cases soft tissues are preserved. About 700 species have been described from the deposits, which give a glimpse of life in and around a 150 million year old coral lagoon.

There is no direct evidence of the shores and of the open seas where the coral reefs flourished but there is indirect evidence that the waters were not under the influence of the tides. The quiet lagoon with its fine limey muds was very much a death trap. All kinds of animals from a wide variety of habitats are preserved. There are no footprints of land animals and the only trackways seem to be of animals in the throes of death.

There are fossils of kingcrabs, found at the end of their own trackways, as in their last moments they staggered round in circles. Perfectly preserved shrimps, crabs and lobsters such as ***Aeger*** (*illustrated*) reveal all the delicate detail of their appendages.

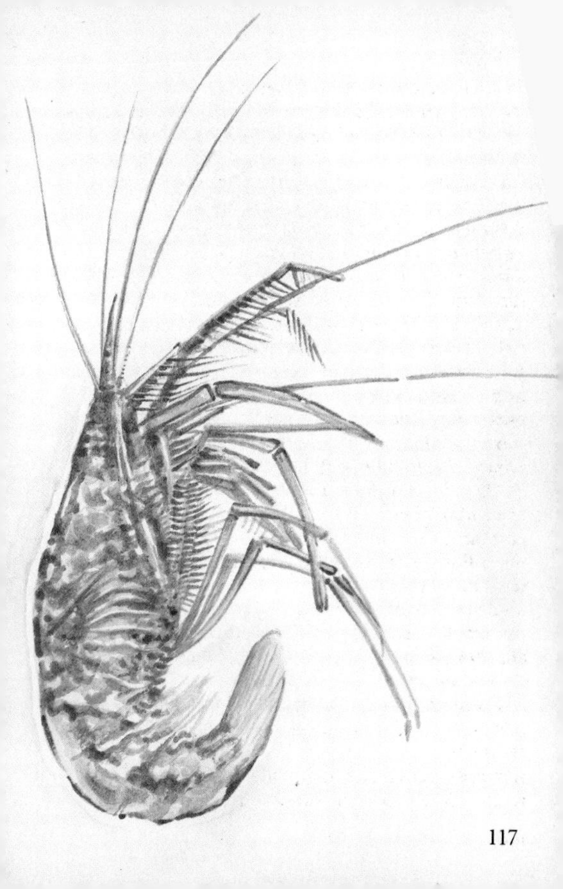

arine worms, with all their tiny limbs perfectly reserved, also coiled up as they died. Flying insects, such as dragonflies and the lacewing *Kalligramma* with its large eye spots (*see page 133*), fell into the water and sank to the bottom to become perfectly preserved. Living in and close to the edge of the lagoon were many kinds of vertebrates.

A number of sharks and several skates occur in the limestone, but the commonest fishes were shoals of small bony sprats, *Leptolepis*. Some fish had long pointed snouts, others were deep bodied and were shellfish crushers or coral crunchers.

Among the reptiles there were several kinds of turtle, inhabiting the shallow waters. Turtles with their bony carapace have hardly changed since they first appeared in the Triassic period. Originally land living, they colonised the seas culminating in the giant 2.5m (8ft) long ***Sokotochelys*** (**1**) from Nigeria and the marine turtle ***Trionyx*** (**2**), which later changed its habitat to become a freshwater pond turtle.

There were five kinds of crocodile, which fed on the 32 different fish genera known from the lagoon. Closer to the shores a number of extremely rare reptiles seem to have accidentally fallen in and been carried out on the tide to drown. This lizard group is of great importance because all the main groupings of living forms are represented. *Bavarisaurus*, a long-tailed gecko, is preserved in the stomach of the tiny dinosaur *Compsognathus*, and there were primitive iguanids, lacertilians, skinks and slow-worms.

1
2

The first bird: *Archaeopteryx*

The most valuable fossils of all time are the six specimens of the earliest-known bird ***Archaeopteryx*** (*illustrated*), from the Jurassic lithographic limestone. The first was bought by the British Museum in 1861, and comprised a small reptile skeleton bearing feathers on the forelimbs and long bony tail. In 1877 a more complete skeleton was discovered with the skull showing a typical reptilian head with teeth.

A third specimen was found in 1951 and another one actually collected in 1855 was not correctly identified until 1970. The fifth a smaller individual found in 1956 was described in 1974, the sixth came to light in 1988.

The importance of *Archaeopteryx* is that it demonstrates how the birds originated from reptiles. The skeleton is reptilian, but it is the possession of feathers that makes *Archaeopteryx* the perfect missing link between birds and reptiles.

Some anti-evolutionists have tried to show that the fossils were faked but the fine details and especially the unique tail feathers and details of the cracks in the limestone slabs have proved that the fossils are genuine.

Archaeopteryx was not a good flier but it could climb trees well and run around on the ground just like a tiny dinosaur. Although, as fliers, less efficient than the furry pterosaurs, the birds had a great advantage when grounded. In addition, feathers are not so vulnerable as a membrane of skin, especially in deep undergrowth.

The Maidenhair Tree

In the Jurassic sandstones of the Yorkshire coast the commonest fossils are leaves of the maidenhair tree, or ***Ginkgo*** (*illustrated*). These trees belong to the gymnosperms or naked seeds and are closely related to firs and pines, and reached up to 30m (98ft) in height. There were male trees which bore catkins and the spores were shed from them and carried by the wind to the female trees. The seeds were encased in a fleshy covering looking a little like a yellow cherry, but which is highly poisonous to man. The most obvious and distinctive part of the *Ginko* is the leaves. They are fan-shaped and often partly divided to give a bi-lobed appearance, hence the name of the only living form, *Ginko biloba*. Fine riblets radiate from the base of the leaves.

During the Jurassic ginkos flourished throughout Europe, Asia, western North America and Greenland. By the Tertiary, 60 million years ago, their distribution was restricted to eastern Asia, the north western part of America in the area of Alaska, Greenland and western Europe. In Pliocene times the ginko was restricted to central Europe and Japan. Today this tree occurs naturally only in parts of eastern China, but has been reintroduced to Europe and America as an ornamental tree.

The Mesozoic Era: The Cretaceous Period (135-65 million years ago)

Cretaceous meadows

The Cretaceous period, 135-65 million years ago, was the time when the ancient continents broke up to produce the modern ones. In the early part of the period it seemed as if life was hardly changing at all. In southern England and western Europe there were extensive flat plains across which shallow, multi-channelled streams meandered. The landscape was covered in vast fields of horsetails.

In the streams there were fish, turtles and crocodiles. There were herds of the plant-eating dinosaurs ***Iguanodon*** (*illustrated*). this ornithopod was one of the first dinosaurs to be recognised and was correctly described by Gideon Mantell in 1825 as a giant plant-eating reptile.

The teeth looked like gigantic versions of the teeth of the living *Iguauna* lizard, hence the name which means *Iguana* tooth. A bony spike was also discovered and as the living *Iguana* has a spike on its snout this is where the bone was originally placed. When complete skeletons were discovered in Belgium in 1877, it was realised it was a spiked thumb.

The numerous worn down teeth of *Iguanadon* indicate that their main diet was tough horsetails. The horsetails are known as scouring rushes and were once used to clean out cooking utensils.

'Claws'

In 1983 in Surrey, England, a huge dinosaur claw (**3**) was discovered in Cretaceous rocks and in 1986 the key parts of the skeleton were described as ***Baryonyx walkeri*** (= Walker's heavy claw, **1**) in honour of the discoverer. The skull was long and narrow and there were twice as many teeth as in the ordinary flesh-eating dinosaurs. The first suggestion was that 'Claws', as it became popularly known, used the large claw on its hands to gaff fish out of streams, it was also thought that the long narrow skull was similar to those of fish-eating crocodiles.

In 1987 a different interpretation was put forward suggesting that the claw was for disembowelling such

dinosaurs as ***Iguandon*** (**2**) and the specialisation of the skull was so that they could thrust their heads into the body cavity to drag out the entrails – 'Claws' was seen to be a specialised visceral feeder.

A number of partly digested fish scales suggest that *Baryonyx* did actually eat fish, but it seems unlikely that fish could have made up the main diet for such a large animal. The development of a single large claw would be more valuable for ripping open other dinosaurs than catching fish. Whatever the answer, *Baryonyx* is a completely new type of flesh-eating dinosaur, belonging to a new dinosaurian family.

Chalk seas

Towards the end of the age of dinosaurs, the seas spread over many parts of the world. There was more area under water than at an other time. There were few mountain ranges being worn away, the land was generally flat and the seas shallow, warm and clear.

The most abundant living things were algae – single-celled plants – which secreted a skeleton of numerous microscopically small oval plates known as **coccoliths** (**1**, greatly magnified) which together cover the spherical alga (**2**, an articulated group of plates). When an individual alga died the small plates would sink to the bottom of the sea and become separated. These microscopic plates of calcium carbonate, invisible to the naked eyes, accumulated in astronomical numbers and they formed the chalk found throughout the world but most familiar as the Downs and chalk cliffs on both sides of the English Channel.

Living on the sea floor were many kinds of sponges which had skeletons made up of tiny spikes or spicules of silica. On the death of the sponges these spicules would dissolve and form jelly-like patches on the sea floor. These silica gels would then eventually solidify and become flints. Those that formed around sponges would form hollow flints and contain fossils of all kinds of microscopic animals.

Along the North Norfolk coast there are huge rings of flint up to 2m (80in) in diameter with, in the very centre, a minute brown ring about 3mm (⅛in) in diameter. These flints are known as 'paramoudra' and

2

formed around some type of worm with a vertical tube up to 2m long.

Sea serpents

With the spread of the chalk seas, a new group of reptiles, the mosasaurs, evolved. They originated from the varanid or monitor lizards but they became enormously elongated and their legs were reduced to paddles. They had long jaws with massive short stubby teeth. They were classic sea serpents. The first skull of a *Mosasaurus* was discovered in Holland in 1780. In 1795 when Napoleon's army bombarded the city of Maastricht they spared the house where they believed it was kept, but, when they got there, the fossil had gone. However, by offering a reward of 600 bottles of wine it was suddenly found! The fossil was taken to Paris where it can be seen to this day.

Since that time mosasaurs have been found in North America and Africa including ***Goronyosaurus*** (*illustrated*) from Nigeria. Their main food seems to have been fish and different types of cephalopods. There is a famous ammonite shell that is perforated by many circular holes. In fact it has been calculated that that particular ammonite was bitten 16 times by a young and presumably inexperienced mosasaur.

Towards the very end of the Cretaceous some mosasaurs became primarily shellfish-eaters and developed almost spherical crushing teeth. These mosasaurs are called *Globidens* (= globe tooth).

At the end of the Cretaceous the ammonites and all the marine reptiles became extinct.

1
4
3
5

Cretaceous insects

The most abundant and prolific animals on the Earth are insects, 75% of all animal species in fact. The fossil record of insects is very patchy. Only certain relatively rare conditions favour their preservation. They may also be missed because people who do not expect to find fossil insects do not look at rocks sufficiently carefully. A dramatic exception to this came in 1984 when hundreds of new species of fossil insects were discovered from the Cretaceous of south east England. There was a comprehensive sample of insect life just immediately prior to the origin of flowering plants.

They included the perfect remains of **dragonflies** (**1**) and damsel flies, which hunt other insects on the wing, and lots of **cockroaches** (**2**). There were the very first termites *Valditermes*, with the line of the wing where the termites shed their wings after their mating flight, and **crickets** (**3**) which were clearly musical as the stridulatory files on the forewing for making the characteristic sounds were preserved. Many bugs were found, mainly leaf hoppers but also cicadas and even a few aphids. **Lacewings** (**4**), with huge eye spots for frightening off potential predators such as lizards, scorpion flies, **beetles** (**5**), diptera or true flies, wood-boring beetles, caddis flies and wasps were all in evidence. The significance of this discovery is that these insects give a clear impression of the range of insect life immediately before the origin of flowering plants, which produced a fundamental change in the life of insects.

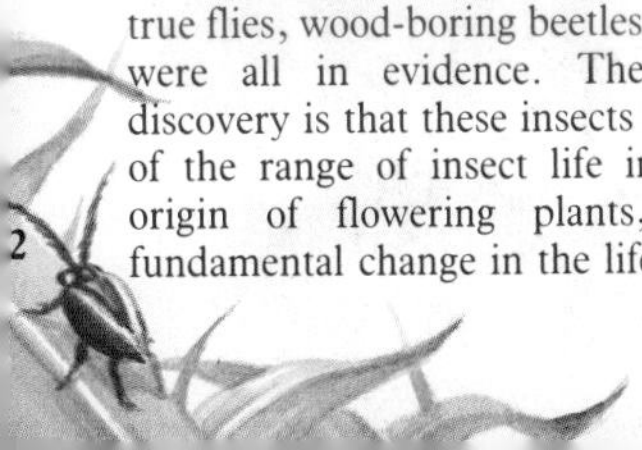

2

The first flowers

During the Cretaceous one of the major events in the evolution of plant life took place: the coming of flowers, which was directly related to insect life. Flowers are special reproductive structures designed to attract insects, which then carry pollen directly from one plant to another, instead of having to rely on chance pollination by the wind. Their component parts are shown here on ***Vahlia saxifraga*** (**1**), a relative of the currants and gooseberries. The *pistil* in the centre of the flower is the female structure which comprises a *stigma* at the end of a slender *style* and a

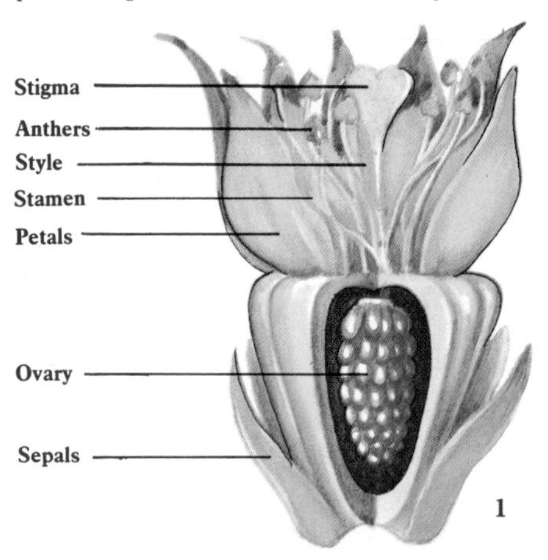

basal *ovary*. *Stamens* surround the pistil and end in *anthers* which carry the pollen, the male element; *petals* surrounding these organs form the *corolla* and the leaf-like *sepals* the *calyx* or cup. Insects, attracted by the flowers and by nectar, collect pollen on their bodies and then deposit it on other flowers which they visit, fertilising them.

Most plants living today, some 220,000 species, are flowering plants. The origin of flowering plants is still not fully understood.

In 1981 Annie Skarby was the first to discover flowers preserved in three dimension, which she named ***Scandianthus*** (**2**), belonging to the saxifrage family. Their sepals, petals, ovary and ovules are all perfectly preserved in exquisite detail. A forest fire had swept over them, turning everything to charcoal instead of reducing it to ash. Charcoal retains fine detail without disintegration.

The armoured dinosaurs

The change in plant life and the spread of the seas were important changes for the dinosaurs. In the northern continents many new types evolved.

Some of the plant-eaters, the ankylosaurs (= fused lizards), which seemed to have descended from the early Jurassic *Scelidosaurus*, developed great thickened bony plates over their bodies, which fused into solid plates in many forms, hence the name of the group. In some the armour formed vertical spikes, in others, such as ***Palaeoscincus*** (= ancient skink, *illustrated*), there were sharp rows of spikes that stuck out sideways.

These ankylosaurs were rather squat but some such as *Nodosaurus* were simply covered in small rectangular bony plates almost like chain mail. The ankylosaurs developed the most complete bony secondary plate of all the dinosaurs. Some of the earlier forms, like *Polacanthus*, described by W. Blows in 1987 from the Isle of Wight, had a kind of bony sheet over the hip girdle and triangular plates in the neck and shoulder region with further sharp vertical plates over the tail.

Many of these dinosaurs weighed about three tonnes and their mere weight ensured their safety as no flesh-eater had the strength to turn them over on their backs. They have been likened to tanks and must have been impregnable.

The ankylosaurs were the only dinosaurs that developed a complete bony palate separating the food and air passages, exactly as found today in mammals.

Ankylosaurs are rare and their remains are often water-worn. It is believed that they lived in dry lowlands and not in swampy lowlands as did so many other dinosaurs.

Bone heads

A group of dinosaurs known as the bone-heads arose from a small 1m (39in) long herbivorous ornithopod from the Isle of Wight, *Yaverlandia*, which had the bone on the top of its head thickened. A later form, *Stegoceras*, from Canada 2m (6½ft) in length, had a dome-shaped head with solid compact bone about 15cm (6in) thick. The bone on an even larger form, ***Pachycephalosaurus*** (= thick-headed lizard, *illustrated*), was thickened to 30cm (12in) with a rim of knobs around the back of the skull and various knobs and spikes on the snout and above the eyes.

An apparently protective thickening on the very top of the head could have only one use – as a kind of battering ram when the animal lowered its head. If two boneheads did lower their heads and butt each other, the force of the impact would be transmitted down the backbone. Head butting – seen in sheep and goats today – is an effective means of defence against enemies. However, it seems likely that this inferred behaviour was comparable to that of sheep and goats and was mainly for trials of strength among the boneheads themselves. If, like most ornithopods, these dinosaurs lived in herds, a dominant male probably won leadership of the herd by combat competition with his fellow males.

Horned dinosaurs

During the late Cretaceous, in the small continent comprising eastern Asia and western North America, a new group of dinosaurs evolved from small ornithopods. These were the ceratopians or horned dinosaurs.

The most primitive type was *Psittacosaurus* (= parrot-reptile) with its hooked beak, the next stage was the 2m (6½t) long *Protoceratops* (*see page 150*). It had a broad bony frill which acted as a support for powerful slicing teeth which chopped up tough plants rather like a straw·cutter. The bony frill made the animal look larger to scare off enemies as well as protecting its vulnerable neck.

Other forms developed spikes and horns as well as frills. The largest and most successful of the frilled lizards was ***Triceratops*** (*illustrated*), 9m (30ft) long. Its frill and horns have generally been taken as a plant-eater's defence against flesh-eaters and it has frequently been portrayed battling with *Tyrannosaurus rex* (*see page 148*), the largest of all dinosaurs, and winning by impaling the flesh-eater on its horns. Healed wounds have been found on fossils of *Triceratops* bony frills but they were actually made by the horns of other *Triceratops*.

The horned dinosaurs' pattern of head decoration can be compared to that in cattle and antelopes. Although the horns would be useful weapons of defence, it seems that their primary purpose was for trials of strength among their own kind. The other feature is that, when the heads are lowered for the

horns to engage, the front view of the frill is very dramatic and may well have been to intimidate rivals or to frighten off potential enemies.

A whole variety of head pieces was developed from the basic types. ***Styracosaurus*** (**4**) had great backward pointing spikes along the posterior margin of the frill. ***Centrosaurus*** (**2**) had two curved processes near the edge of the frill that looked like handles for holding on to, and could only have been for decoration. ***Chasmosaurus*** (**1**) had a straight margin to the back of the frill, while ***Pentaceratops*** (**3**) had small triangular plates running around the entire margin of the frill.

All the ceratopians had elaborate frills and a pattern of horns but they were not solid bony frills with the exception of *Triceratops*, in which the frill is formed on a solid sheet of bone.

The inferred pattern of behaviour suggests that the ceratopians lived in herds. In the Dinosaur Provincial Park in Alberta, Canada, there is a deposit composed entirely of the remains of the ceratopian *Centrosaurus* and this occurrence has been interpreted as part of a herd that had drowned while attempting to cross a river. Dinosaurs of all kinds are scattered throughout the Dinosaur Park, there is just this one area packed with dozens of *Centrosaurus*, which is strong evidence that as well as dying together they must have been together in life. The ceratopians were among the most successful plant-eaters towards the end of the Cretaceous and they evolved a large number of different types.

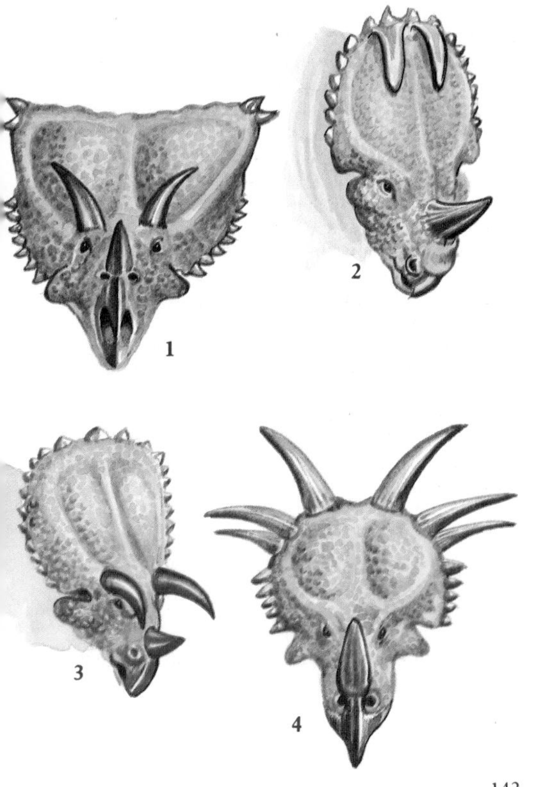
1
2
3
4

Duck-billed dinosaurs

The most successful of all the herbivorous dinosaurs were the hadrosaurs or duck-billed dinosaurs. These were large ornithopods with tails flattened from side to side and strengthened by ossified tendons. The front feet were webbed. Although they seemed adapted for life in the water, fossilised stomach contents showed that they fed on pine needles and cones. Of all dinosaurs, the hadrosaurs were best able to feed on tough plant materials; they had as many as 2000 teeth in their jaws which were able to grind up tough plants. The teeth formed several closely packed parallel rows, each with a layer of hard enamel along the outer edge. This arrangement made an ideal rasping surface for grinding up plants. The duck-bill was used to collect food and muscular cheeks helped in the chewing process.

A striking feature of hadrosaurs was the variety of head shapes, many developed enormous crests, some projecting 1m (39in) behind the skull.

In the main the crests were the result of the increase in the size of the nasal bone and the premaxilla (bone at the front of the upper jaw). In ***Corythosaurus*** (*illustrated*) the rounded helmet is made up of nasal bone. The crests enclosed a complex folding of the nasal passages. This may have given the hadrosaurs a more acute sense of smell to warn them of enemies downwind or the cavities may have served as resonating chambers enabling the dinosaurs to call to each other over large distances with great grunts and bellows.

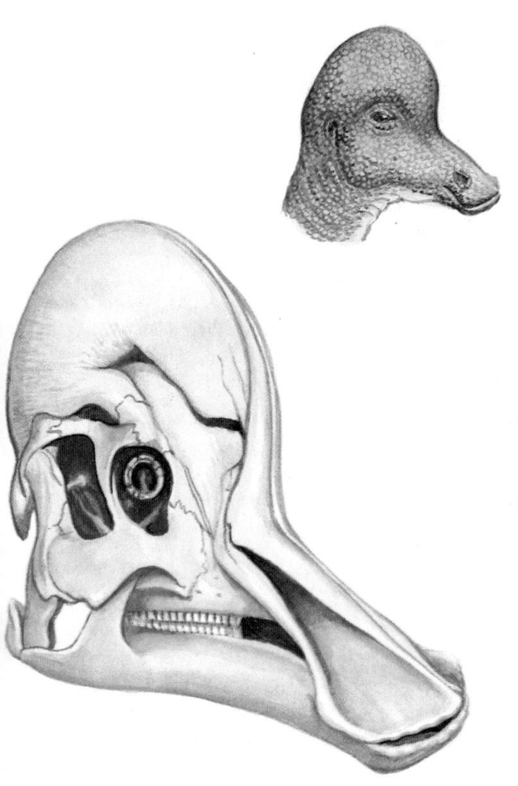

Dinosaur nurseries

A discovery, made by J. Horner in Montana in 1979, completely changed everyone's ideas about the way of life of dinosaurs. A circular nest 3m (10ft) in diameter and standing 1.5m (5ft) high had a saucer-like hollow 2m (6½ft) across and 75cm (30in) deep excavated out of the summit. In the hollow were preserved the remains of 11 duck-billed dinosaurs. There were four other individuals just outside the nest.

Unlike earlier discoveries in Asia, where dinosaur nests and hatchlings have been found, the young were about 1m (39in) long and already their teeth were well worn with feeding on tough plants. The occurrence of 15 young dinosaurs in a hollow at the top of a

prominent mound, and exposed to the attentions of any passing flesh-eater, seems to be extremely unlikely. The only explanation of this situation is that the young must have been protected by the adults. From this evidence it is now accepted ***Maiasaura*** (= mother lizard, *illustrated*) and other hadrosaurs looked after their young. It is imagined that like many birds, such as ostriches and geese, the young would follow their mother when she went foraging for food.

It is now known that crocodiles build similar mound-like nests in which to lay their eggs and on hatching the mother looks after her young. Recently, in Montana, it was discovered that the hadrosaurs had large communal nesting sites and nurseries for bringing up their young.

Tyrannosaurus rex (*illustrated*)

Tyrannosaurus rex (= king of the tyrant lizards) was the largest and fiercest looking flesh-eater of all time, with a length of 12m (40ft) and weight of 6 tonnes. *Tyrannosaurus* walked on its hind legs with the backbone and tail held horizontally and the head raised up on a flexible neck like a swan or goose. Fossil footprints show that *Tyrannosaurus* waddled along just like a goose, pigeon-toed and with its thick tail swinging from side to side with each step, at about 4kph (2½mph) or perhaps 6kph when in a hurry. This means that *Tyrannosaurus* was probably a scavenger feeding on dead and dying dinosaurs and was not an active hunter as portrayed in popular films.

The 15cm (6in) long, flat, blade-like teeth had serrated edges, just like steak knives, ideally suited for dealing with inert meat but not for living flesh.

The tiny two-fingered hands could have been used to pick rotting flesh from between the teeth. Their main function was to make it possible for *Tyrannosaurus* to stand up after it had been laying down. When a sitting *Tyrannosaurus* straightened its folded up hind limbs its body would simply have slid along the ground. This action was prevented by the two hands sticking in the ground, so that the force could be transmitted into lifting up the body.

For its size *Tyrannosaurus* did not require much food. One duck-billed dinosaur could provide the equivalent of one year's energy needs, a brontosaur three, a *Brachiosaurus* ten times as much.

A fight to the death

The main active hunters among the dinosaurs were the clawed dinosaurs, such as *Deinonychus* (= terrible claw, *illustrated on the title page*), which stood 2m (6½ft) tall and 3m (10ft) in length. They hunted in packs and attacked their victims with a specialised claw on the hind feet which was enlarged into a sickle-like slashing organ. The tail was rigid, being encased by all the tail vertebrae having elongated extensions, so that the tail was strengthened by some 40 rods of bone. The tail acted as a balancing organ, enabling *Deinonychus* to stand on one foot while ripping open its prey, as well as leaping on to its victims.

In relation to their body size, the brains of the clawed dinosaurs were about 25 times better than brontosaurs' and six times than that of *Tyrannosaurus*. They were about as brainy as an ostrich. The main development was in the parts concerned with muscular coordination and balance.

The only direct evidence of dinosaurs fighting came to light in Mongolia in 1971. Two complete dinosaur skeletons were found entwined together. A young plant-eating ***Protoceratops*** (**1**) lies on its side with its sharp beak in the chest of a young lightly built clawed ***Velociraptor*** (= speedy robber, **2**) which has its sickle claw and hands wrapped around the head of the *Protoceratops*. Both inexperienced individuals died in the encounter.

This discovery is astonishing because the two animals must have been almost immediately buried, perhaps by a sandstorm, so that the two corpses remained undisturbed until their discovery in 1971.

The discovery of the clawed dinosaurs changed many old ideas about dinosaurs because this group was fast moving and extremely agile. The slow, sluggish life of the dinosaur did not apply to these swift, pack-hunting killers.

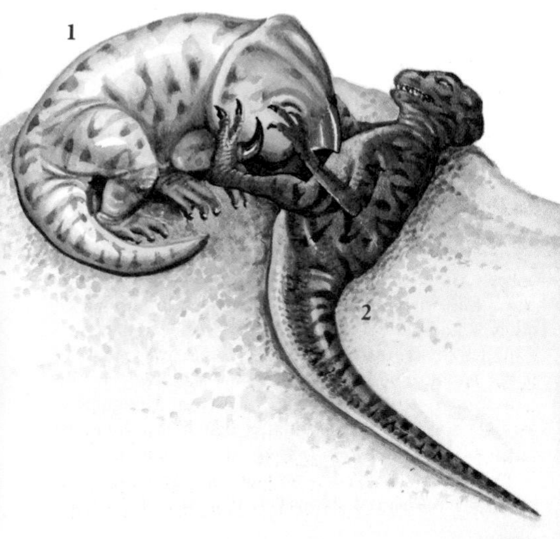

Ostrich dinosaurs

Towards the end of the Cretaceous period there evolved a group of lightly-built, long-legged but toothless carnivores. They had long, upright necks and small heads with a large beak. They held their backs horizontally and, although they had long straight tails, were otherwise very ostrich-like in their proportions – hence the name ostrich dinosaurs. *Struthiomimus* (= ostrich mimic) was found in Alberta, Canada. One skeleton *Oviraptor* (= egg robber) from Mongolia was found on top of a nest of *Protoceratops* eggs, and this suggested that the ostrich dinosaurs may have eaten the eggs of other dinosaurs.

Dromiceiomimus, from Alberta, had the largest eyes of any land-living animal and, like other ostrich dinosaurs, was able to focus both eyes on the same object and hence see it in three dimensions. With the long fingers of their grasping hands and long necks, they would have been capable of accurately snatching small animals, such as mammals, that came out as the light faded. The largest, 4m (13ft) long ***Gallimimus*** (*illustrated*) was described in 1972 from Mongolia.

One scientist has speculated that if the dinosaurs had not died out then perhaps the ostrich dinosaurs could have evolved into an intelligent bipedal reptilian humanoid. However, the parts of the brain that were developed were for muscular coordination not intelligence, so the nearest they would have got would have been level with the ostriches.

A new dinosaur invention

In 1980 a new dinosaur was described from Salta province of Argentina and named ***Saltasaurus*** (*illustrated*). A type of brontosaur, 12m (40ft) long, it was among the last of the sauropods which were first known from rocks of late Triassic age and flourished especially during the Jurassic. Relatively few sauropods continued through to the late Cretaceous and these were mainly in the southern continents.

Saltasaurus, unlike any others of its kind, possessed a bony armour. Embedded in the skin over the back and halfway down the sides of the body were irregular bony plates, about 12cm (5in) in diameter, set in a background mass of tiny bony nodules 5mm (⅕in) across. This type of armour is not found in any other dinosaur, but the great suprise was for a sauropod to have any armour at all.

The basic pattern of armoured dinosaurs is derived from the rows of rectangular bony plates, or scutes, just beneath the skin, as in modern crocodiles. This regular arrangement of plates and/or spikes is then developed into different arrangements as in the stegosaurs (*page 138*) or the ankylosaurs (*page 136*). In *Saltasaurus* the pattern is irregular, a new invention among dinosaur armour, not derived from the original arrangement. One group of mammals, the armadillos, also re-invented a bony armour, seen in its most dramatic expression in *Glyptodon* (*page 194*), with bony plates also set in a background of bony studs, in much the same way as in *Saltasaurus*.

Cretaceous toothed birds

In 1880 O.C. Marsh, the chief palaeontologist of the US Geological Survey, published a 100-page monograph on fossil birds from the Chalk of Kansas. Charles Darwin, just before he died, wrote to Marsh that 'these old birds had afforded the best support for the theory of evolution, which has appeared within the last 20 years.' Marsh was later forced to resign his job because of pressure from anti-evolutionist politicians: 'Birds with teeth. That's where your hard-earned money goes, folks, on some professor's silly bird with teeth.'

The significance of these birds is that they were intermediate between the first Jurassic bird *Archaeopteryx* (*see page 120*) and modern toothless beaked birds. These Cretaceous birds had lost the long bony reptile tail but still retained their reptilian teeth.

Icthyornis (= fish bird, **1**) was a tern-sized seabird, and like all modern birds no longer had a bony tail but still bore teeth.

Another kind of toothed bird ***Hesperornis*** (= western bird, **2**) was even more remarkable as its wings were tiny and hence this bird was flightless. Its hind legs were powerful and the main means of swimming and diving. The bones of the skeleton were not filled with air, as in living birds, so this would have helped in diving. A study of the toes shows that they were not web-footed but had leaf-like expansions of the skin. They share many features with loons and grebes but were not directly related, just an example of animals developing to fit particular niches.

1
2

The rise of birds

During the Cretaceous there was a gradual increase in bird life, at the expense of the pterosaurs. The birds had two advantages over the pterosaurs. When grounded they could tuck their wings out of the way and run around like small bipedal dinosaurs, while pterosaurs could only manage an awkward waddle. If the feathered wing of birds is torn by dense undergrowth it will simply part and lost feathers will grow again but a tear in a pterosaur membrane is certain death.

One of the first environments to be colonised by birds was that of the shores and estuaries. Running along the edges of the water, feeding on insects and worms as well as plants and berries, were small

rail-like birds such as ***Palaeotringa*** (**3**) which would have probed wet mud for worms, arthropods and shellfish. In shallow lakes ***Gallornis*** (**2**) would have filtered algae from the water in the manner of flamingos. ***Parascaniornis*** (**1**) was an ibis- or goose-like form and with its long legs probably hunted fish and amphibians in shallow water.

At sea the toothed birds and the cormorant-like ***Elopteryx*** (**4**) competed directly with the pterosaurs, but the birds had the advantage of being able to dive. Although many of these modern-looking birds fit well-recognized niches it is not possible to be certain that they are directly related to their living counterparts, It is more likely that they are examples of convergent evolution.

Quetzelcoatlus

Before the end of the age of dinosaurs, nearly all the pterosaurs had died out. Only the giant *Pteranodon*, with its 8m (26ft) wingspan, continued to soar over the oceans, swooping up fish in its large pouched bill. A large, delicate and paper-thin crest projecting behind its head served as a controlling rudder.

It was believed that *Pteranodon* had reached the maximum size possible for a flying animal. In 1975 the first remains were discovered in Texas of a much larger pterosaur: ***Quetzelcoatlus*** (*illustrated*), named after the Central American feathered serpent god Quetzalcoatl. It had a 10m (40ft) wingspan, its neck was very long and there was a long narrow bill. This is now recognized as belonging to a new family of pterosaurs, the azhdarchids, a name which comes from the Uzbeck word for dragon.

Recent studies of material from Kazakhstan have shown that *Quetzelcoatlus* had a neck which could only move up and down and was specialised for dipping down into the water to snatch fish. Originally it was thought that *Quetzelcoatlus* was a vulture-like scavenger, soaring over the plains and feasting on dead dinosaur corpses. The neck vertebrae's restricted movement made such a mode of life impossible.

Discoveries in petrified forests, once like mangrove swamps at the edge of the sea, recently revealed remains of baby pterosaurs around the upright trunks, suggesting that pterosaurs nested up in the trees and fledglings often fell out of their nests to drown in the tidal lagoons.

Night mammals
The earliest known true mammals, from the end of the Triassic 190 million years ago, were found in South Wales, China, North America and Lesotho in southern Africa. Throughout the period of the age of dinosaurs until 65 million years ago they occupied the niche of the night. At first they were insect eaters with pointed cusps on their cheek teeth for cracking the carapaces of beetles.

One group of mammals developed chisel-like incisors and filled the gnawers and nibblers niche now occupied by the rodents. A number of plant-eaters lived in the undergrowth and were the ancestors of most of the later herbivorous mammals.

The marsupials or pouched mammals were represented by the opossums which feed on most kinds of food, both animal and plant. The other generalised mammals were insectivores and these seem to have been the ancestors of both the modern insectivores and the carnivores. Towards the end of the Cretaceous, 65 million years ago, the first primate ***Purgatorius*** (*illustrated*) occurs, named after Purgatory Hill in Montana. It is known from the crowns of molar teeth and is almost indistinguishable from the living tree-shrew, the most primitive member of the Primates, the mammalian order to which we belong.

The warm-blooded mammals were active during the night when similar-sized reptiles went into torpor. The only enemies of the night mammals were the ostrich dinosaurs which hunted them as the light faded.

The extinction of the dinosaurs: bang or whimper?
At the end of the Cretaceous period, 65 million years ago, the dinosaurs died out. The reason for this mass extinction still baffles scientists. However in 1980 Luis Alvarez believed he had solved the mystery. The end of the age of dinosaurs was marked in the rocks by a layer of clay enriched by a number of rare metals related to platinum, in particular iridium. The concentration of iridium seemed to be similar to that found in meteorites. Alvarez then calculated that such a clay deposit could have been produced by a meteorite about 15km in diameter striking the Earth. Further calculations suggested that, whether the meteorite landed in the seas or on land, the dust

resulting from the impact could have cut out the sunlight for about three or four years.

The consequences of this is that plant life would have died off so that the herbivores would have starved to death and the carnivores that preyed on them would also have died out. In this way the giant dinosaurs were wiped out, but small mammals and birds and in fact most animals less than 60kg in weight could have managed to survive this crisis.

There is no evidence of meteoric impact on land and so it was assumed it must have crashed into the ocean, producing gigantic tidal waves that would have swept over all the low lying continents, causing utter devastation. There is now some evidence that there

were wild fires across the world and this is thought to have been a further consequence of the meteoric impact.

At the end of the Cretaceous many groups of organisms became extinct: microscopic animals and plants inhabiting the surface waters of the oceans, the ammonites, the plesiosaurs and mosasaurs in the seas, the giant pterosaurs in the skies, and on land many types of plant and primitive mammal as well as the dinosaurs.

From careful dating it is established that the different groups did not all die out at the same time, but there was up to 100,000 years separating some of the extinctions.

At the same time there were other groups that were not affected at all: birds in the air, lizards, turtles, crocodiles, insectivore and herbivorous mammals and flowering plants on land, and squids and cuttlefish and fishes in the seas.

The dinosaurs began to go into a decline five million years before the end of the Cretaceous. About 300,000 years before the end this decline accelerated and at the very end there were only twelve species left which belonged to only eight groups. They were: **ankylosaurs** (**1**), ***Tyrannosaurus*** (**2**), **hypsilophodont** (**3**), **bonehead** (**4**), **duckbill** (**5**), **ostrich dinosaur** (**6**), ***Triceratops*** (**7**) and a **clawed dinosaur** (**8**).

The great extinction was not a sudden violent event but took place gradually over several million years. The iridium-rich clay was most likely from volcanic eruptions.

1
2
3
4
5
6
7
8

The Tertiary Era
(65-1.6 million years ago)

The dawn of the Age of Mammals

With the final disappearance of the dinosaurs the mammals were presented with a world of unoccupied environments just waiting to be colonised. During the Palaeocene (65-5.3 million years ago) there was a rapid evolution of the mammals to fill these empty niches. The primitive herbivores which included ***Meniscotherium*** (**2**), had claws and a heavy tail and these gave rise to several types of specialised forms such as the squat heavily built *Coryphodon* which had hippopotamus-like proportions, and the running *Phenacodus*. Many mammals remained in the forests living in the undergrowth and up in the trees, one form *Planetotherium* became a glider like the living flying lemur. Small long slender carnivores, the hyaenodonts, evolved from the primitive insect-eating mammals, and a dog-like group, the mesonychids, arose from the primitive herbivores.

Carnivorous birds

There were no large predators among the mammals, this role was taken by giant flightless flesh-eating birds. They were about 2m (6½ft) tall with heads the size of a modern horse and huge hooked beaks as well as strong claws on their feet. ***Diatryma*** (**1**) was the main predator at the beginning of the age of mammals. In South America, which was cut off from

1
2

the rest of the world, the main flesh-eater was another giant flightless bird, *Phorusrhacos*. It was quite unrelated, had a more hooked beak and with longer and thinner legs was a faster runner.

The tropical forests of Europe

By Eocene times, 53-34 million years ago, the giant flesh-eating birds had been replaced by mammalian carnivores. Most life on land was in woodland and forest and the greater part of Europe was clothed in tropical forests, similar to those of Asia. In the Thames estuary the London Clay contains many

fossils of plants, fruits and leaves, indeed J.Parsons in 1757 believed that the biblical Flood took place in the autumn on the evidence of these fossil fruits. Mangroves, with their prop roots, and palms were the dominant plants but on the drier parts there were many types of familiar plants. ***Uvaria*** (**1**), the sweet sop now found in Africa and tropical Asia, ***Magnolia*** (**2**), the surviving most primitive of the flowering plants, ***Hibbertia*** (**3**), now found only in Madagascar and Australasia, the ragged rose ***Oncobia*** (**4**), the blackberry ***Rubus*** (**5**), belonging to the rose family, and the fern-like ***Sabal*** (**6**) as well as many trees such as oak, ash, poplar, sycamore, willow, maple and birch, breadfruit and fig trees.

At Messel, in Germany, there is a unique lake deposit in which flowers, leaves, fruits and pollen are all preserved in exquisite detail. Over 50 kinds of flowers have been found, apart from one catkin from a walnut tree all the flowers were insect pollinated. Laurel, witch hazel, myrtle, silkwood, dogwood, vines, periwinkles and oleander flourished around the margins of this lake.

Messel

The Eocene lake deposit of Messel, in Germany, is one of the most important fossil localities in the world because it contains an important sample of the ancestors of many groups of living mammals. However, what makes Messel unique is the way in which the remains have been preserved. Complete and undisturbed skeletons of birds and bats have been excavated in considerable numbers, but what is sensational is that in the birds the entire plumage of feathers is intact and in the case of the bats the wing membranes and even the large ears are preserved in all their details. Far and away the most abundant fossils are those of flying animals: insects, birds and bats. It looks as if gases such as carbon dioxide were given off and anything flying into them would be immediately killed and fall into the lake. The feathers and skin are actually preserved as films of bacteria.

As well as the unique preservation, Messel is remarkable for its rare fossil mammals which include the earliest records of many groups.

Animals living near this poisonous lake would occasionally be gassed and become fossilised. In 1978 the first known pangolin ***Eomanis*** (**1**) was found and in 1981 the first anteater ***Eurotamandua*** (**2**) which, until this discovery, was a type of mammal known only from South Africa. The animals were lying on their sides, the skeletons perfectly preserved with all the bones articulated.

Horses which must have come to the water's edge, there to become asphyxiated and collapse, are among

1
2

the more spectacular fossils from Messel (**1**). There is one perfect skeleton that has a fully formed foal still preserved within the mother. The outline of the body is clearly visible and in some fossils the horse's mane is present, running along the midline from the region of the eye to just above the shoulder. The thin tail had a bristly brush-like end, quite different from the way in which their tails had previously been imagined.

This early horse, called ***Propalaeotherium*** (**2**), stood about 40cm (16in) high and lived in forests. From the structure of their feet, with four toes on the forelimbs and only three on the hind, they were adapted to walking and running on soft ground.

The rounded cusps on the cheek teeth have led scientists to conclude that the early horses were browsers, feeding on leaves. The Messel horses have their stomach contents preserved. They are a mass of plant debris but, when carefully extracted and examined under the microscope, it is possible to see the minute holes for respiration, the stomata, that are found on the underside of leaves. Messel provides proof of the diet of the early horses.

The evolution of the horse

The fossil history of the horse is one of the classic documents of evolution. From the dog-sized, four-toed browsing *Hyracotherium* from the Eocene London Clay and *Propalaeotherium* from Messel to the modern living horse *Equus* the main changes that took place were a gradual increase in size and a

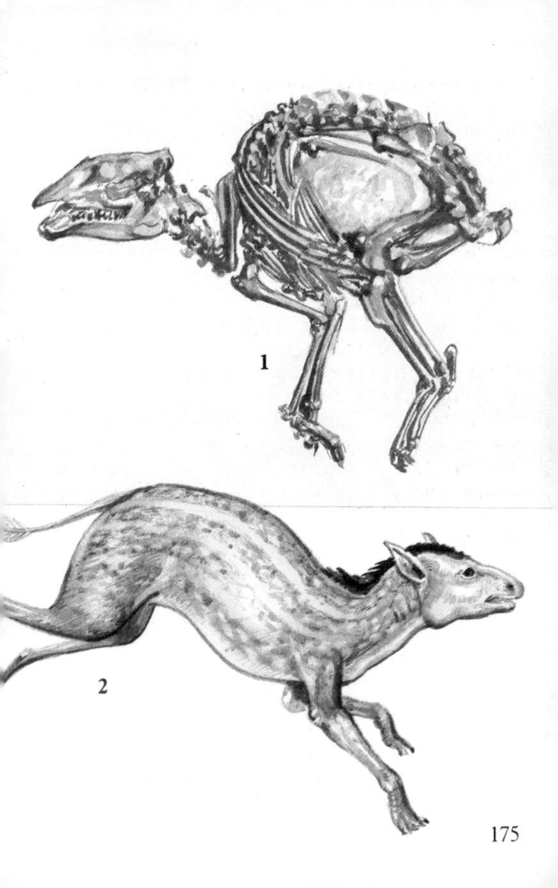
1
2

reduction of the bones between the ankles and wrists and the toes. The legs became comparatively long and the advanced horses ran on the tips of their toes. The teeth changed from having only a few rounded cusps to having a chewing surface of complex enamel ridges and long open roots that allowed them to continue growing.

These developments were connected with the ability to run at speed over firm ground and a change

in diet from browsing on leaves to grazing on grass. The newly evolved grasses first appeared in the Oligocene (34-23 million years ago) but had spread worldwide by the Miocene (25-5.3 million years ago). Grasses contain silica, the substance of sand and glass, which wears down teeth very effectively.

At the end of the last century it was discovered that the horses had evolved in North America and that there had been a series of migrations from America to Asia and Europe across the Bering Land Bridge. The main stages in horse evolution include the Eocene ***Hyraotherium*** (**1**), Oligocene ***Mesohippus*** (**2**), and Miocene ***Merychippus*** (**3**). All these were leaf-eaters. A major change took place with the development of the Pliocene (5.3-1.6 million years ago) grass-eater ***Pliohippus*** (**4**).

Chalicotheres

During the Eocene and Oligocene, from 53 to 23 million years ago, there were three main groups of odd-toed hooved animals: horses, tapirs and rhinoceroses. All three evolved specialised running forms, with similar proportions. From the horse line, however, there arose the chalicotheres (= small stone beasts) which had such a peculiar mixture of features that they almost sabotaged the study of fossil vertebrates. Baron Georges Cuvier (1769-1832), the founder of the subject, proposed a fundamental law of correlation of parts, in which he insisted that hooves would always be associated with the grinding teeth of a plant-eater and sharp claws with the shearing teeth of a meat-eater.

The chalicotheres, such as ***Chalicotherium*** (*illustrated*), had an horse-like head but had feet bearing three large claws, an impossible combination of characters according to Cuvier. The sharp claws of the hands could be swung in a wide arc and they were probably turned under, as these animals seem to have walked on their knuckles. No-one understands what they were specialised for, but they survived for about 50 million years, the last dying out in Africa about one million years ago. An animal that survived so comparatively recently overlaps with early man.

In South America a different group of mammal, the clawed toxodont *Homalodotherium*, developed all the main chalicothere features quite independently and was also adapted to the same specialised way of life – even though we do not know what that was.

Rhinoceroses

At the beginning of the age of mammals, the odd-toed hooved mammals were the important herbivores. The early horses, tapirs and rhinoceroses were all small forest dwelling browsers and, apart from small differences in their teeth, all looked much the same. The early rhinoceros were known as the running rhinos, for they had thin spindly legs. One line of rhinos became very hippo-like, these were the amynodonts that were eventually replaced by true hippopotamuses.

In Asia, during the Oligocene (34-23 million years ago), there evolved a line of gigantic hornless rhinos. Remains of these giants are found from the Caucasus across central Asia to China. ***Indricotherium*** (*illustrated*) was the largest land mammal ever known, standing 5.5m (18ft) at the shoulder and weighing 30 tonnes, three times as much as the sauropod dinosaur *Diplodocus*.

During the Miocene (23-5.3 million years ago), the heavily built rhinoceroses spread over North America, Asia and Africa and many developed horns of matted hair. These rhino horns were never preserved but roughened raised bony lumps on the snouts show where they were formed. Today only five species survive but during the Pleistocene ice ages the long-haired woolly rhino *Coelodonta* flourished and in the southern Russian *Elasmotherium* there was a form with a 2m (6½) long horn. Studies on bones of the woolly rhino show the original proteins still preserved intact.

The intelligent hunters

One of the major changes during the age of mammals was the spread of grasslands during the Miocene (23-5.3 million years ago). Horses and antelopes all became adapted to feeding on grasses and also became fleet of foot. With open grassland the flesh-eaters could not easily approach their prey, as they could be seen over large distances. The only way the flesh-eaters could succeed was by using their intelligence. From small polecat-like forms living in the trees, two distinctive hunting strategies developed for catching prey over open exposed grasslands.

Stabbing and biting cats

The cats developed cunning and stealth, carefuly stalking their victims and pouncing on them. The cats are the most highly specialised flesh-eaters with blade-like shearing teeth, the carnassials, which act as scissors to chop the meat off the bone. They also have powerful forelimbs for bringing their victims to the ground.

There were two different techniques for despatching prey, biting the neck as living cats do and

stabbing with greatly elongated knife-like canines. The stabbing cats, or sabre-tooths, attacked large slow-moving animals such as elephants and mammoths. The first cats appeared in the Oligocene (34-23 million years ago). Among them were sabre-tooths such as ***Eusmilus*** (**1**) and the biting cat ***Nimravus*** (**2**). Fossils have provided evidence of a fight between these species in which the stabbing teeth of *Eusmilus* damaged the frontal sinuses of *Nimravus*, though the wound subsequently healed.

Running Dogs

The second group of carnivores that succeeded in hunting in open grassland were the dogs. The dogs became specialised for running on their toes. The bones of their forelimbs were fused so that they could not rotate, in contrast to cats which have very flexible forelimbs (watch a cat with a ball of wool or a mouse). This rigidity is better adapted for running on firm ground. Skeletons of the modern dog *Canis* are known from Pliocene rocks of Europe, but the dogs first arose in the Oligocene of North America.

The success of dogs is due to their high level of intelligence, which in their case led to organised hunting by a pack under a leader. Their intelligence is also recognisable in their social nature and the way in which they can adapt to other intelligent social hunting animals, such as man.

A dog pack selects a particular member of a herd of herbivores, often one that seems to be a likely victim, a slower individual or lame animal. They cut it out from the herd then drive it to and fro until it weakens. Then the dogs close in and pull their victim to the ground, tearing at its underbelly. It is the

weight of several dogs combined that overcomes the prey. The disembowelling results in a rapid death.

During the Miocene a group of heavily built, hyaena-like scavenging dogs developed – the osteobores. Bears evolved from dogs in Europe and later spread to the Americas. But there were earlier animals in North America, the amphicyonids or bear-dogs, which had dog-like heads and bear-like bodies. The bears gave rise to the giant panda and another group, the racoons, to a prehistoric giant South American 'panda'.

Whales

After the extinction of the marine reptiles such as the mosasaurs, ichthyosaurs and plesiosaurs at the end of the age of dinosaurs, the mammals returned to the sea. The first fossil whale comes from the beginning of the Eocene, 53 million years ago. This is ***Pakicetus*** (*illustrated*) from Pakistan. The teeth show that the whales were descended from the dog-like mesonychid condylarths. During the Eocene there were two contrasting types: the 20m (65ft) long *Basilosaurus*, which had a small head and probably fed on squid, and the small 2.5m (8ft) dolphin-shaped *Pappocetus* which was a fish-eater.

Basilosaurus had the overall proportions of mosasaurs and *Pappocetus* that of icthyosaurs. The modern toothed whales and dolphins evolved from these ancient whales, and the first whale with whalebone, or baleen, instead of teeth, *Mauicetus*, is known from the Oligocene of New Zealand. The baleen, sheets of mineralised hairs suspended from the roof of the mouth, filtered krill, shrimp-like crustaceans, from the waters. The baleen whales culminated in the 130 tonne Blue whale.

Other mammals returned to the sea, from the dog group came the eared seals or sealions and the shellfish-eating walruses and from the polecat group the Miocene *Potomotherium* gave rise to the true seals.

Giant deer

Towards the end of the Oligocene, 23 million years ago, the odd-toed hooved mammals, the horses and rhinos not withstanding, were in decline and the even-toed were in the ascendant. The more primitive forms were the pigs, which gave rise to the hippopotamuses. the advanced forms were the ruminants or cud-chewers. They developed a new system for dealing with tough plant materials. From their teeth they do not appear to be well adapted for dealing with grasses. However, they developed a separate stomach, or rumen, in which bacteria and single-celled organisms mixed with mucus break down plant material. This is then regurgitated as cud and rechewed and swallowed down into the normal stomach.

The deer, which first appeared in the Miocene, developed bony antlers, for trials of strength between males, but shed and regrew them every year. The largest antlers come from the Irish elk ***Megaloceros*** (*illustrated*) and had a 3.7m (12ft) wide span and probably weighed 45kg (99lbs). For many years it was believed that these antlers were so large that they could only have been used as display structures to intimidate rivals. In 1987 a study of the microstructure of the bone proved that they had been used in fighting.

Deer and Pronghorns

The main evolution of deer took place in North America. There was a great variety of styles of head decoration. ***Procranioceras*** (**1**) had three small branched antlers on the top of a bony pedicle; ***Syndyoceras*** (**4**) belonged to the most primitive group of deer and had four unbranched antlers.

The major change in the history of the ruminants was when a horny sheath, of the same material as hooves and nails, was formed over the bony outgrowths. The living pronghorn of North America has a horny covering which is shed each year and during the Miocene there was an enormous variety of pronghorns, some, such as ***Hexameryx*** (**3**) from Florida, had six horns.

In the Miocene of Italy a small ruminant appeared with long, stabbing canine teeth, ***Hoplitomeryx*** (**2**). It had five bony outgrowths. It is not known whether these were covered by horn or shed annually like antlers. In France, Africa and Mongolia there are bony horn cores of gazelle-like antelopes, which had horny sheaths that were never shed. These were the first bovids, the most advanced of all the ruminants and the ancestors of cattle and sheep. In Africa and Asia a great variety of forms evolved, and is still represented on the grasslands of East Africa by the many kinds of antelope, such as hartebeests, oryxes and duikers as well as sheep and cattle.

The herbivores were highly specialised in their food preferences; the conditions in East Africa today represent a kind of fossil ecosystem.

1
2
3
4

Bison

About five million years ago the bovids underwent a sudden evolutionary expansion: 50 genera known from Asia, 20 from Europe and 12 from Africa. As well as the familiar antelopes of East Africa, the ancestor of the modern cattle, the nilgai which still survives in India, and the ancestor of sheep and goats, the chamois, all flourished.

At the end of the Pleistocene (1.6 million years ago) cattle, bighorn sheep and musk ox crossed into North America over the Bering Land Bridge from Asia. The cattle that reached the plains of North America were ***Bison antiquus*** (**1**) and the less common long-horned ***Bison latifrons*** (**2**).

The bighorn sheep and sole surviving pronghorn occupied the more mountainous parts and the musk ox and moose the cold arctic region.

The bison came to dominate the North American plains to the virtual exclusion of all other hooved herbivores, even the horse that had evolved for 50 million years in North America finally died out.

A new evolutionary strategy developed: now just a few highly adaptable forms dominated. Instead of having highly specific types of food, they were able to eat almost any kind of plants. This was especially important because the world climate was changing and getting colder. The seasonal changes in plant life meant that herbivores could not afford to be fussy about their food preferences. It was the unspecialised feeders that came into their own: the cattle, sheep and goats.

1
2

South American mammals

For most of the age of mammals, South America was cut off from the rest of the world. There were a number of primitive groups of mammal, some of which are known from the Eocene of Europe in Messel, and they gave rise to many different types inhabiting particular environments and looking almost exactly the same as the quite unrelated mammals that lived in similar conditions in other parts of the world.

There were three kinds of mammal that seemed to be unique to South America: the anteaters, sloths and armadillos. Giant ground sloths (*see page 228*) lived until 11,000 years ago. But when North and South America became reconnected during the Pliocene (5.3-1.6 million years ago) the ground sloths migrated north as far as Alaska. The first remains from North America were described as *Megalonyx* (= giant claw) by Thomas Jefferson in 1799. The first fossil anteater comes from Europe (*see pages 172-3*) and the first armadillo *Utaetus* with hinged bony plates from North America.

A major branch of the armadillos culminated in the 3.3m (11ft) long and 1.5m (5ft) high armoured ***Glyptodon*** (*illustrated*) which grazed the grasslands of South America and later reached North America. The rigid bony carapace served as a protection with the tail a formidable weapon of defence.

It was the clear relationship of the ground sloths to the living tree sloths and glyptodonts to armadillos that first led Charles Darwin to realise that the living forms were related to the different extinct types and that therefore the species must have changed over millions of years, that evolution must have taken place.

The convergent evolution of mammals

The isolation of South America from the rest of the world allowed the mammals to conduct a kind of experiment in evolution that could be compared with what was going on in the rest of the world.

From the primitive herbivore animals there developed a fast running camel or llama-like group and a small horse-like group, which culminated in the one-toed ***Thoatherium*** (**1**). A rhinoceros-like group, the toxodonts, included both the heavily built and small running forms, just as in other parts of the world. There were even highly specialised clawed toxodonts which filled the niche of the chalicotheres (*see page 178*).

The astrapotheres lived the lives of hippopotamuses and the pyrotheres seemed to be elephant-like. There were rodent-like typotheres and rabbit-like hegetotheres. Although they looked similar to mammals from other parts of the world, they were

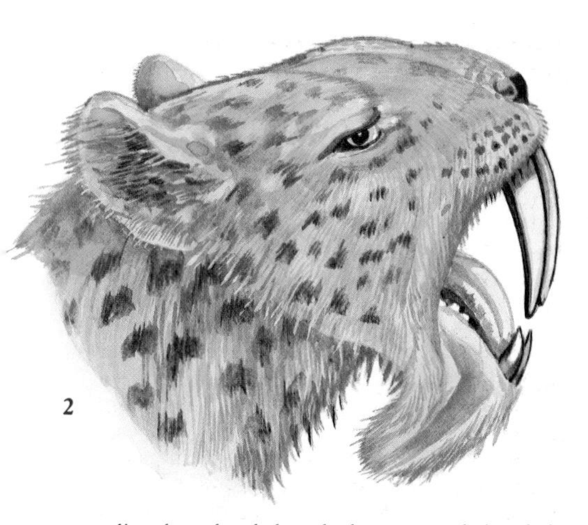

not directly related but had converged in their structures, as they adapted to comparable ways of life.

The only carnivores in South America were marsupials or pouched mammals. The caenolestids were the main insect eaters and looked like shrews, these still survive in the Andes mountains. There were also hyaena-like forms but the most dramatic was the marsupial stabbing cat ***Thylacosmilus*** (**2**) from the Miocene and Pliocene of Patagonia. This pouched sabre-tooth preyed on the large ground sloths and rhinoceros-like toxodonts. It filled the ecological niche of the sabre-toothed tigers exactly.

Egg-laying mammals of Australia

The most primitive mammals are the egg-laying spiny anteaters and the duck-billed platypus, which also produce milk for their offspring. In many ways they are intermediate between reptiles and mammals. They lay eggs identical to those of birds and reptiles. However, they have hair and can produce milk. This is exuded by specialised glands in the skin. No breasts or mammae are formed, nor special structures such as nipples for suckling. The young simply suck droplets of milk from the skin. These mammals are only known from Australasia, and until very recently there was no hint of their existence in the fossil record.

In 1984 remains of an egg-laying mammal were discovered in New South Wales from 100 million year old Cretaceous rocks. Until then the first record had been in 1975 when ***Obdurodon*** (**1**) a tooth-bearing duckbilled platypus was described from South

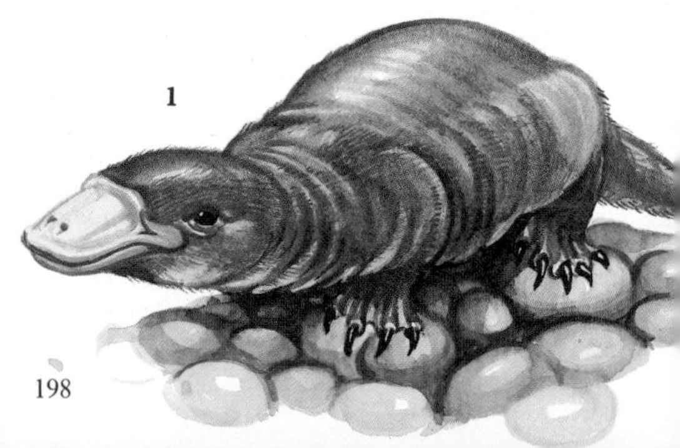

Australia in 15 million year old Miocene rocks. The living platypus has a longer bill but no teeth although in the young teeth do form but they are rapidly shed.

The living echidnas feed on ants and termites but the mountain form ***Zaglossus*** (**3**) which lives in New Guinea mainly eats worms from the leaf litter of the forest floor, **Giant *Zaglossus*** (**2**) have been found in New South Wales and Western Australia from the Pliocene (5.3-1.6 million years ago) and surviving up to 20,000 years ago in Tasmania. From their remains it is clear that they could eat larger items of food than ants, and with their powerful forelimbs could have dug for beetles, grubs and worms in the ground and in rotten logs.

Marsupial Lion *Thylacoleo*

At the beginning of the age of mammals fragments of a marsupial or pouched mammal were preserved on the continent of Antarctica. Young marsupials are born as tiny almost worm-like creatures which crawl up into a pouch (Latin: *marsupium*) where they attach to a nipple from which they obtain their milk. It is believed that the marsupials reached Australia from South America via the Antarctic before the Antarctic ice cap began to form during the Oligocene (34-23 million years ago). In contrast to South America the marsupials filled all the major niches. The marsupial mice were actually insect-eaters, and the Tasmanian Devil and its relatives lived in forest and woodland and were the 'cats' of Australia.

Towards the end of the Miocene, with the spread of grasslands, the thylacines or Tasmanian 'wolves' evolved; they had the proportions of wolves and were adapted for running down their prey. The numbat or banded anteater and the marsupial mole are two further examples of convergent evolution.

Thylacoleo (*illustrated*) was one of the most puzzling marsupials. It was the size of a leopard and had stabbing incisor teeth at the front, behind which was a pair of long, blade-like teeth for slicing its food. The skeleton shows that it was related to the fruit- and flower-eating possums or phalangers, which include the sugar glider. Some authors thought at first that *Thylacoleo* may have been a fruit-eater but fine scratches on the teeth prove it was a flesh-eater.

Giant kangaroos and wombats

The first plant-eating marsupial *Wynyardia* comes from Oligocene rocks in Tasmania and is the first diprodont (= two front teeth). By Miocene times the three living families were established: the possums or phalangers, the wombats and the kangaroos. The first phalanger was a primitive koala bear and further tree-living types gave rise to forms which, like South American monkeys, were able to use their tails like a fifth hand, a gliding flying lemur-like form, as well as a flesh-eater, the cuscus.

With the spread of grasslands the main herbivores were the wallabies and kangaroos, which filled the niche of the ungulates. They were unique in running by hopping on their hind legs. The largest kangaroo was the 3m (10ft) tall ***Procoptodon*** (**1**) which had a short, rounded face and short tail. The third group included the wombats, which were specialised for burrowing, but these gave rise to huge 4m (13ft) long forms with the overall proportions of rhinoceroses which were probably grazers. One form, *Palorchestes*, had long sharp claws and probably a short trunk, a sort of mixture of chalicothere and elephant. The largest true wombat ***Phascolonus*** (**2**) inhabited the margins of rivers and lakes and was not a burrower.

In spite of the introduction of placental mammals, in which the early development is within the mother and who exchange food and wastes via a special structure (the placenta), the Australian marsupials have survived to the present day. They are well adapted to the conditions of Australia, although

1
2

some, such as the koala, have become specialised. It feeds only on eucalyptus leaves.

Giant mammals

During the ice ages over the past two million years ice sheets have regularly advanced and retreated over much of North America and Eurasia. The climatic zones of plant life have similarly shifted and so have the mammals. the warm or temperate mammals and the cold-loving forms have migrated to and fro as the climate alternately warmed and cooled.

Throughout this period many of the familiar modern animals, such as horses, bison, oxen, deer, wolves and elephants, were in existence but were significantly larger. In North America, from Florida to Alaska, there were giant beavers ***Casteroides*** (**1**), which had large webbed feet and were 2.75m (9ft) long, weighing about 200kg (440lb). Across Eurasia there roamed the giant hyaena ***Pachycrocuta*** (**2**), 2.5m (8ft) in length. In Siberia the rhinoceros ***Elasmotherium*** (**3**), 6.5m (21ft) in length, was double the size of the largest living rhinoceros.

The development of these giants may be related to the ice ages in that a larger size is more efficient at conserving heat, as the surface area through which the heat is lost is comparatively small relative to volume. The other advantage of large size is that it makes them less vulnerable to the attentions of flesh-eaters. This cannot be the whole story because some mammals, such as the woolly mammoths, seem to have become smaller during the colder periods.

Island pygmies

One of the major consequences of the ice ages is that with the expansion of ice sheets locking water up in vast quantities, the sea level throughout the world fell by about 100m (327ft) and then, when the ice retreated, rose again. Lowlands flooded and animals were trapped in hilly regions which formed small islands.

Large herbivores cut off on islands would give rise to miniature races. The Maldive Islands in the Indian Ocean were inhabited by the dwarf elephants *Stegodon*; off the coast of California there was a race of tiny woolly mammoths. On the islands in the Mediterranean, such as Malta, there was a minature giant deer ***Megaloceras*** (**1**) only 1.5m (5ft) tall. On islands in the Caribbean the giant ground sloth ***Megatheriu*** (**2**), normally 4m (13ft) tall, was the size of a domestic cat. Perhaps the most dramatic dwarfing was the 1m (39in) tall ***Palaeoloxodon*** (**3**) the straight-tusked elephant from Malta.

Whilst larger mammals became smaller and smaller many of the normally small mammals developed giant versions. Rodents, such as dormice and rats, grew to about a quarter of the size of the island elephants. On the Balearic Islands, off the south coast of Spain, a rodent grew to the size and proportions of a goat, but with the chisel incisor teeth of a rodent.

The origin of humans

At the dawn of the age of mammals, 65 million years ago, there already lived up in the trees a small tree

1
2
3

shrew-like mammal such as *Plesiadapis*. The eyes were positioned on the front of the face so that they could judge distances as they jumped about high up in the trees. The thumbs were opposable to all the other fingers, so that they could grasp objects and manipulate them. These features are the trademark of the Primates.

In North America and Europe the Primates became a highly successful group of mammals, and as they became better adapted to life in the trees their faces became flatter, the long snout being reduced as the sense of smell became less important.

Squirrels and Primates

During the Eocene, 53 million years ago, the true gnawing and nibbling mammals, the rodents, such as the early squirrel ***Paramys*** (**1**), began to take over from the Primates, while others, such as the advanced tarsier-like ***Tetonius*** (**2**) with its large eyes were clearly evolving towards a monkey-like condition.

Although a few managed to survive into later geological periods the Primates of the northern hemisphere became extinct.

The Primates continued in South America, where they evolved into the new world monkeys which developed a prehensile tail, which serves as a fifth hand. The main development of the Primates took place in Africa. In Oligocene (34-23 million years old) rocks a tree-living Primate *Aegyptopithecus* (= ape of Egypt) was discovered. It was about the size of a domestic cat, but was the ancestor of both the apes and humans.

African apes *Dryopithecus*

In Africa the forest-dwelling Primates occur in Oligocene rocks in Egypt. During the Miocene period (23-5.3 million years ago) the advanced Primates, such as the ancestral gibbon *Pliopithecus*, swung through the trees by their arms. Also at this time, grasslands spread at the expense of the forests and ***Dryopithecus*** (*illustrated*), a small 1m (39in) high , ape-like Primate, came out of the forests to live in the open. The other significant event during the Miocene was that Africa re-established contact with Eurasia, in the west of Spain and in the Middle East, and this allowed the grassland living dryopithecines to migrate from Africa into Eurasia.

Dryopithecus was able to flourish on the open plains on account of its high intelligence. They had the largest brains for their size of any mammal and furthermore they must have lived in groups – this is inferred because as separate individuals they would have been at the mercy of any predator.

The teeth indicate that the dryopithecines were among the first apes. They walked on all fours but could stand up on their hind legs for a good view of their surroundings and to warn the group of the approach of danger.

Many of the features that are recognized as human are already to be found in these lightly built apes which gave rise to the orang utans of Asia and the chimpanzees and gorillas of Africa. It is from the African branch of the apes that humans originated during the Pliocene (5.3-1.6 million years ago).

Laetoli footprints

The first sign of humans on Earth comes from 3.6 million-year-old rocks from Tanzania. The evidence is not in the form of fossil bones or teeth but in a trail of footprints left by a small human family which was discovered by Mary Leakey in 1979.

In this part of East Africa there are volcanoes which produce carbonatite larva and ash, which is chemically the same as limestone. During a series of eruptions clouds of ash were thrown up and with the accompanying rains the countryside was covered in a layer of wet cement. The nests and eggs of ground-nesting birds such as guineafowl, as well as rodents in their burrows, were killed instantly in the ash fall, but the other birds and mammals were not seriously affected. In fact, they walked across the wet but drying muds leaving perfect footprints. Giraffes, antelopes – in fact virtually every local mammal known – seems to have left its footprints preserved in the wet cement.

Far and away the most significant tracks are those of a 1.3 tall advanced Primate which walked upright on its hind feet – that walked as humans do. From the details of the prints it is concluded that the tracks were made by the first human beings. The larger prints are thought to have been those of an adult male, a smaller set are assumed to have been an adult female who was stepping carefully into the footsteps of the male walking in front. A third set of smaller footprints runs alongside and these were made by a child whose hand was being held by its mother.

The Quaternary Era: The Ice Ages
(1.6 million years ago to the present day)

The age of humans

In 1925 the anthropologist Raymond Dart described the skull of a young ape-like animal from 1.6 million-years old Pleistocene rocks at Taungs, in Botswana, southern Africa. He named this fossil *Australopithecus africanus*, the African southern ape, and he claimed that it represented an intermediate animal on the line to man. The brain was about the size of a chimpanzee's but the separate regions of the brain were human-like, the teeth were human-like and the head was held on a vertical backbone proving that this individual walked upright. At first the experts rejected Dart's views because, from apparent evidence of early man already found, they believed that the large brain came first: Piltdown man had a human brain and ape jaws (they did not know that Piltdown man was a hoax) and Neanderthal man (a possibly divergent form of early human, *see page 224*) had a human skull and jaws but seemed to have walked with an ape-like stoop. Later it was discovered that Neanderthals walked exactly as we do.

Many subsequent discoveries of australopithecines have confirmed Dart's original ideas on human evolution, and the adults were exactly as he had predicted from his study of the skull of the child. The most complete skeleton, known as **Lucy** (*illustrated*), was discovered in 1979 in Ethiopia and exactly such

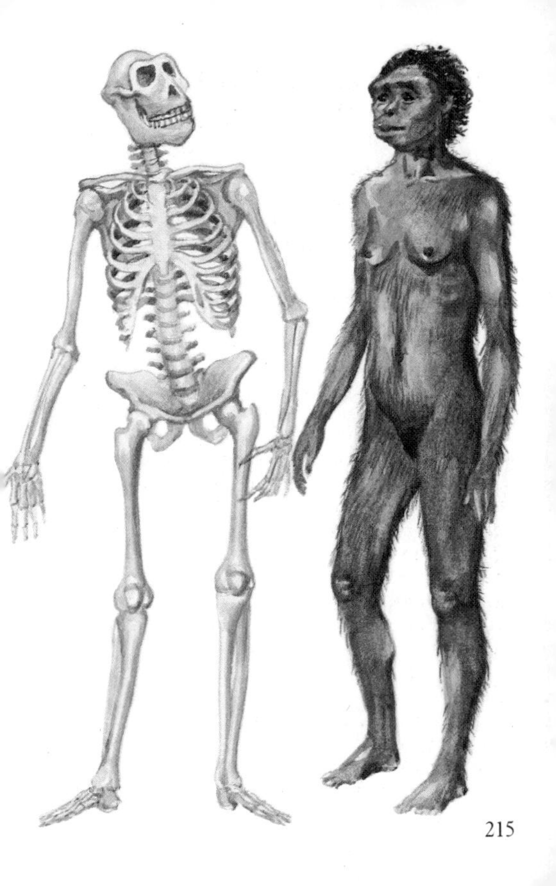

a being could have made the footprints found preserved at Laetoli.

The death of an australopithecine child

The major studies on the lives of the australopithecines were by Raymond Dart and he discovered that one of the major events in human evolution was the change in diet to a more meat-eating way of life and to being a group hunter. Dart showed that the australopithecines used bones, teeth and horns as tools and weapons. He named this the osteodontokeratic (= bone, tooth, horn) culture. In one set of caves baboon skulls were found which had been fractured by having been struck with the ends of antelope limb bones. At the same time simple chipped-pebble tools were also manufactured.

At one site at Swartkrans in the Transvaal, South Africa, there is a fissure filled with the heads, hands and feet of baboons and australopithecines. Today, in limestone country, the edges of such fissures are the only place where trees can get a hold and grow. Here leopards bring the prey they have caught, taking it up a tree to prevent its being stolen by lions or hyaenas. The leopards eat their food up in the trees but the hands, feet and heads of their prey usually drop down. The Swartkrans fossils are almost entirely the remains of leopards' meals.

One of these fossils is the skull of a child (**2**) on which are fractures that exactly match the canine teeth of a leopard. The cat must have carried the child off (**1**) by gripping the head with the upper teeth in the eye sockets and the lower ones in the back of the head.

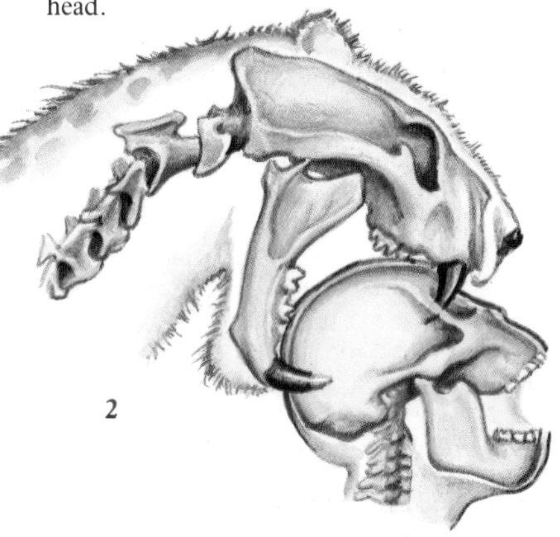

Fire

One of the greatest technological discoveries in the history of humankind was the discovery, use and control of fire. In 1981 the earliest evidence of man-made fire was discovered at Lake Baringo, Kenya, in rocks 1,400,000 years old and associated with simple stone tools. Fire gave humans complete independence from the rest of the animal world – it provided protection against predators as well as warmth. It also meant that meat could be cooked, this made it more digestible and helped preserve. Outweighing all these factors was the use of fire as a hunting tool.

At Terra Amata, in southern France, a hunting band came every spring and built a communal

shelter. There was a hearth for cooking, a place for butchering and they slept on hides laid on the ground.

At Ambrona in central Spain, about 500,000 years ago, the first true human and our immediate forerunner, *Homo erectus* (= upright man), set fire to the grasses each year as the straight-tusked elephants ***Palaeoloxodon*** (illustrated) migrated through the region. The animals were driven into boggy ground and butchered on the spot. Tools were actually made on the site. A number of hunting groups cooperated. The butchering sites all had examples of all the various types of animal killed, there had been an equal sharing out of the hunting spoils, as primitive hunters share today.

Petralona skull from Greece

The first fossil human skull from Greece was found in 1960 in the Petralona cave, and is thought to be about 500,000 years old. The back of the skull is similar to that of *Homo erectus*, but the face and the ridges over the eyes, as well as a brain size of 1200cc (73c in), are similar to *Homo sapiens*. In fact there are six features which link the Petralona skull with *Homo erectus* and eight with *Homo sapiens*.

The Petralona skull has been described by Chris Stringer of the British Museum as sitting on the boundary between advanced types of *H. erectus* and primitive *H. sapiens*. and is important because of its mixture of features. The origin of our own species has been a subject of considerable argument; some authors suggest that we were not directly descended from *Homo erectus*, while others believe that we arose by means of a sudden evolutionary jump from *Homo erectus*.

This fossil from Greece has helped to settle all the arguments – it is a perfect intermediate form and shows that there was a gradual transition from *Homo erectus* to *Homo sapiens*. The recent discovery in Africa of a complete skeleton of *Homo erectus* suggests that modern man's origin was from Africa and not Asia.

Studies of human DNA and blood proteins suggest that mankind branched off from the African apes during the Pliocene (5.3 million years ago) but that modern humans date from only about 200,000 years ago. The divisions into modern races is a very recent

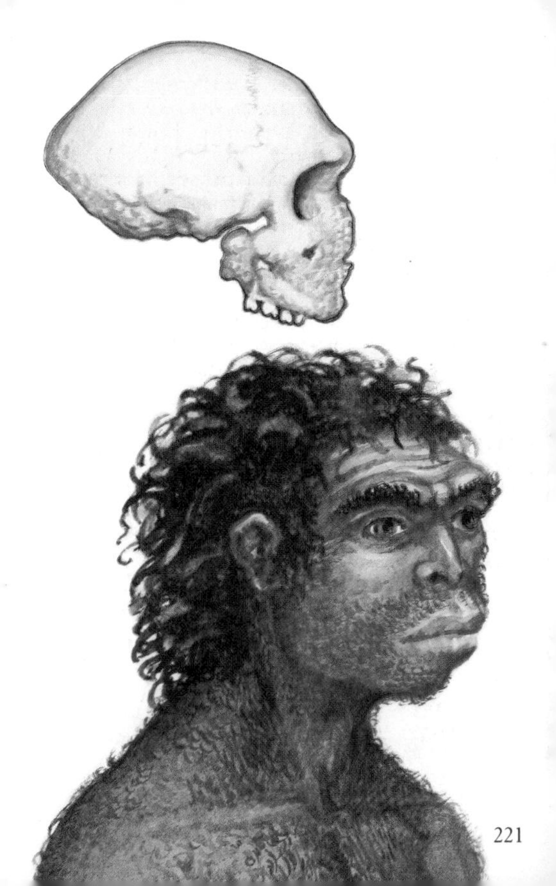

evolutionary event that took place after the retreat of the last ice age, 10,000 years ago.

Beneath Trafalgar Square

Before the last ice sheet advanced, the climate in Europe, 100,000 years ago, was much warmer than it is today. In 1957 there were excavations for Uganda house, in Trafalgar Square in the middle of London. A team of experts on fossil shells, plants and mammals examined the remains that were collected, including 13,000 snails and bivalves and 150 different kinds of plant. Some of the snails lived in dry grassland, others among rushes in marshy ground. There were many types of beetle including the rose chafer which feeds on wild rose and dung beetles as well as ground and water beetles. Along the ancient river Thames, there were reeds and bullrushes, waterchestnut and yellow waterlilies. There were

southern European maple trees, hazel and hawthorn shrubs.

Teeth of the straight-tusked elephant were found, they had first been noted from Pall Mall as long ago as 1731. The large ox, the auroch, *Bos primigenius*, red deer and fallow deer were among the grass eaters, but rhinoceroses were also around. Wallowing in the water were **hippopotamus** (**1**); this was the time when they spread from Africa to northern Europe, as far north as Yorkshire. Bear and hyaenas were present and also the **cave lion** (**2**), which is the most striking coincidence as Landseer's lions dominate the Square, however, from the evidence of cave paintings, the male lions did not have manes.

A Neanderthal burial

The early types of modern humans are generally known as the neanderthals. With their short stature and heavy brow ridges, they are often portrayed as brutish looking, with a shambling ape-like gait. This view is based on a skeleton of a crippled 60-year-old man suffering from severe osteoarthritis and is not at all representative of normal healthy neanderthals.

Neanderthals that were crippled from birth were looked after, as indeed were the sick. They also took especial care in burying their dead.

A remarkable discovery was made recently at Shanidar in Iraq. From the very careful and meticulous excavations, it was found that in the soil covering the body there were separate patches of pollen from particular flowers. This distribution of pollen could only have been achieved if posies of flowers had been placed in the grave prior to burying the corpse. Posies of **groundsel** (**1**), **cornflowers** (**2**), **hollyhocks** (**3**) and **grape hyacinth** (**4**) were left in the grave by the mourners.

This discovery gives an insight into the society of early man and the reverence for life and perhaps a belief in an afterlife – in any event it indicates that man must have acquired some religious sense by this time. This is also indicated by the ritual burial of cave bear skulls deep in caves.

The living quarters were usually near the mouth of the caves, where fires were built and tent-like shelters of animal hides were constructed.

Mammoth hunters

Thirty thousand years ago the neanderthals were replaced in Europe by Cro-magnon man. This was during the last ice age, and in the flat plains of tundra along the margins of the ice sheet roamed herds of woolly mammoths ***Elephas primigenius*** (**2**). The name mammoth comes from a Siberian word for an underground creature. Often, when the ground thawed, mammoths came to light that had been deep frozen in crevasses during the last ice age. The local people believed that they inhabited the underworld and hence their name. Woolly rhinoceroses, musk ox, reindeer and many small animals such as foxes and hares also flourished.

In the Ukraine there developed a mammoth based economy. Mammoths, especially young ones, were the prime object of the hunt. Even the skeletons were put to good use: human dwellings were built of walls

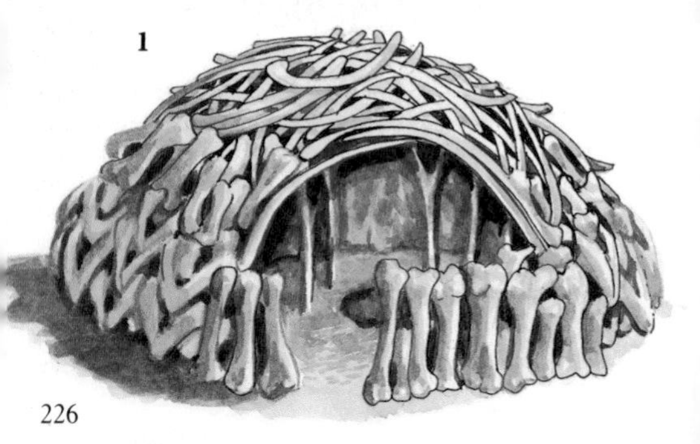

of stacked up mammoth jaws and limb bones, the roofs were constructed of ribs tied in with reindeer antlers. One of these circular huts (**1**) was made of 385 bones from 95 mammoths. The overall covering would have been of mammoth skins. The living area was about 25 sq m (30 sq yd) and in the centre were ashes from their hearths. There is plenty of evidence of habitation with numerous flint implements. There were antler hammers, bone awls and needles, ivory spears and also engravings and carved figurines of mammals and humans, the latter generally female. In 1981 it was discovered that they made musical instruments out of bone. Clothing was of hare skins.

There was plenty of food and, for the first time, abundant evidence of man's artistic creativity.

Rancho La Brea tar pits

At Rancho La Brea in the middle of Los Angeles, in California, is one of the richest of all fossil deposits, full of skeletons dating from 40,000 to 4,000 years ago. Here, over year upon year, crude oil has seeped up to the surface, the petrol evaporating to leave behind sticky pools of tar or asphalt.

In the past animals attracted by the water that would have collected on the surface have become trapped. Between 1913 and 1915 some 750,000 bones were excavated from the Rancho La Brea tar pits. The evidence of man is rare; only a single skeleton of a woman 1.5m (4ft 11in) tall, aged between 20 and 25 years old, from 9,000-years-old tar.

The giant Imperial mammoth, the small American mastodon, bison, horse, camel and giant ground sloth, ***Megatherium*** (**1**), are all typical plant-eaters that ended up in the tar. There are occasionally true wolves, Californian lions, pumas, bobcats and foxes, but it is sabre-tooth tigers and heavily-built, hyaena-like dire wolves, ***Canis dirus*** (**2**), as well as the Western vulture, ***Coragypus*** (**3**), that are present in enormous numbers. About 90% of the species are flesh-eaters – they normally make up about 3%. This unusual deposit is dominated by animals selected by their stupidity; the more intelligent animals, such as the Californian lions and wolves, recognized the dangers of the tar pits and carefully avoided them.

3
2
1

Cave art

The most common remains of prehistoric man's activities are stone tools. The **hand axes** (**1**) and **flakes** (**2**) found in Swanscombe in Kent, in south-east England, were the basic tools of *Homo erectus*. Subsequently, with modern humans, *Homo sapiens*, very fine flint implements were made for specific tasks such as **daggers** (**3**), and **tanged arrowheads** (**4**), by the technique of pressure flaking. These more delicately manufactured tools were the result of humans developing the precision grip, where implements are held between the thumb and forefingers. This ability distinguishes *Homo sapiens* from his forebears.

However, 30,000 years ago, a further important change took place and that was the representation of animals in the form of **cave paintings** (**5**), as at

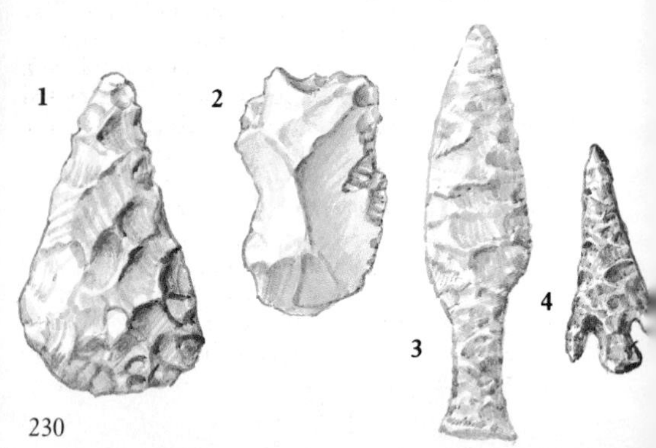

Lascaux in France, but also in three-dimensional sculptures and engravings on stone and ivory. The actual appearance in life of the giant Irsh elk and the woolly mammoth and rhino is only known with certainty from the detailed and accurate paintings and etchings. Portrayals of human beings were also common, mainly of females but not exclusively so. The details of the human form are remarkable in that they portray the entire range of the shapes of the human body accurately, without any concessions to fashion. Different hair styles and personal ornaments are carefully documented. There is no way of knowing the purpose of this cave art, many of the animals portrayed were not actually the main animals that were hunted.

The Neolithic revolution

After the retreat of the last ice age, 10,000 years ago, a further major event in the evolution of humankind took place: the Neolithic revolution. Instead of hunting animals they were herded, in fact domesticated: sheep in Iraq 10,500 years ago, goats in Iran 9,500, pigs in Turkey 9,000 and cattle in Greece 8,500 years ago. The other neolithic innovation was the husbanding of grain, when ground down it could be made into a nutritious porridge and when left to ferment it produced beer.

The first stage of this major change in the lifestyle of humans comes from Jarmo in Iraq, where there was a hunter-gatherer community which had settled – there were 24 mud houses – and begun herding sheep and harvesting grain. By about 8,000 years ago, at **Catal Huyuk** (**1**) in south-central Turkey, for example, there were established small towns of mud houses.

As well as the domestication of animals and plants, a whole variety of new skills had developed: the

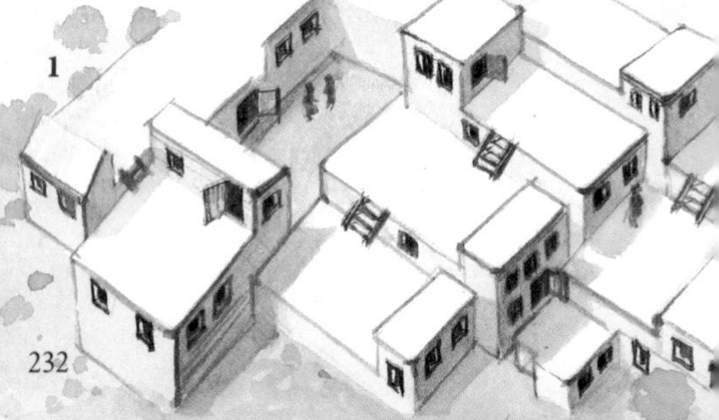
1

weaving of baskets and matting, the firing of pots, the making of bricks, the weaving of cloth, the baking of bread and the brewing of beer (**2**).

With the development of peaceful crafts and the beginning of trade the Neolithic heralds the beginning of civilisation. The word itself is derived from the Latin word *civitatem*, for town.

Index